Root Biology - Growth, Physiology, and Functions

Edited by Takuji Ohyama

Published in London, United Kingdom

IntechOpen

Supporting open minds since 2005

Root Biology - Growth, Physiology, and Functions
http://dx.doi.org/10.5772/intechopen.77875
Edited by Takuji Ohyama

Contributors
Abha Mishra, Yoshihiro Kobae, Martino Schillaci, Sneha Gupta, Robert Walker, Ute Roessner, Salwa Ahmed Latif, Hanan Abou Zeid, Christine Stöhr, Hannah Holzgreve, Manuela Eick, Hiroki Matsuoka, Kei Kumakura, Taito Kobayashi, Wataru Kobayashi, Asaka Takahashi

First published in London, United Kingdom, 2019 by IntechOpen
IntechOpen is the global imprint of INTECHOPEN LIMITED, registered in England and Wales, registration number: 11086078, 7th floor, 10 Lower Thames Street, London, EC3R 6AF, United Kingdom
Printed in Croatia

British Library Cataloguing-in-Publication Data
A catalogue record for this book is available from the British Library

Additional hard and PDF copies can be obtained from orders@intechopen.com

Root Biology - Growth, Physiology, and Functions
Edited by Takuji Ohyama
p. cm.
Print ISBN 978-1-78985-309-4
Online ISBN 978-1-78985-310-0
eBook (PDF) ISBN 978-1-83962-839-9

For EU product safety concerns:
IN TECH d.o.o., Prolaz Marije Krucifikse Kozulić 3, 51000 Rijeka, Croatia, info@intechopen.com or visit our website at intechopen.com.

Meet the editor

Takuji Ohyama has been a Professor in the Department of Agricultural Chemistry, Tokyo University of Agriculture since 2017. He obtained a Ph.D. degree from the University of Tokyo in 1980, and he was a Professor of Niigata University from 1982 to 2017. He has served as a Dean of the Graduate School of Science and Technology at Niigata University (2010-2014). He was a president of the Japanese Society of Soil Science and Plant Nutrition from 2007 to 2009. His research interests are: nitrogen fixation and metabolism in soybean plants, new technology of deep placement of slow release nitrogen fertilizers for soybean cultivation, and nitrogen and carbon metabolism in tulip, curcuma, sugarcane, rice, and cucumber plants. He is also interested in the use of stable isotopes.

Contents

Preface

People love beautiful flowers; however, people usually don't care about the roots hidden in the soil. Most plants have roots, which anchor the plant in soil and physically support the above-ground part of the plant. In addition, roots absorb water and nutrients from the soil and transport this to the shoot. In return, roots obtain photoassimilates from the shoots to support the root growth and function. Although plants have a wide variety of above-ground parts with respect to size, color, and shape of the flowers, leaves, and stems, the primary structure of the root is surprisingly similar among plant species. The root initiates from a seed, root, or stem, and the root grows by cell proliferation in the meristem tissue in the root tip, and the cells differentiate into the epidermis, cortex, and stele. Water and nutrients are absorbed through the cell membrane of epidermis and cortex via transporters or channels. The water and nutrients are transported to the above-ground parts of the plant via xylem vessels. The root growth and functions are affected by various abiotic and biotic environmental conditions, such as levels of water, salt, acid stresses, and the presence of soil diseases. However, beneficial microorganisms such as rhizobia, mycorrhizal fungi, and some microbes in the rhizosphere help plant growth and nutrient absorption. This book intends to provide some up-to-date knowledge of root biology.

This book describes new aspects of root biology related to growth, physiology, and functions. There are three sections in this book, 1) Root Structure and Function, 2) Plant Microbe Interactions, and 3) Metabolites and Human Health.

In the first section, protease activity in the rhizosphere of tomato plants was observed by rhizoboxes in relation to plant growth parameter and nitrogen supply (Chapter 1). Chapter 2 presents the morphological and physiological root plasticity in rice investigated by the relationship with water regime and soil bacterial densities. Chapter 3 describes the analysis of nitrogen transport in barley by tracer experiments using positron emitting 13N and stable 15N labeled nitrate.

In the second section, Chapter 4 studies the importance of plant growth-promoting rhizobacteria (PGPR) and the main mechanisms of the interaction between PGPR and plants. Chapter 5 looks at the new findings of the short life span of the infection unit of arbuscular mycorrhizal fungi (AMF) colonized in the host roots.

In the third section, Chapter 6 describes the changes in the concentration of metabolites in salted white radish root, a traditional Japanese food, during food processing.

I gratefully acknowledge all of the authors of the interesting chapters in this book. I also thank Mr. Josip Knapić, an author service manager of IntechOpen, for his sincere assistance in promoting, editing, and publishing this book.

Dr. Takuji Ohyama
Department of Agricultural Chemistry,
Faculty of Applied Biosciences,
Tokyo University of Agriculture,
Tokyo, Japan

Section 1

Root Structure and Function

Chapter 1

Protease Activity in the Rhizosphere of Tomato Plants Is Independent from Nitrogen Status

Hannah Holzgreve, Manuela Eick and Christine Stöhr

Abstract

Rhizoboxes were developed in order to analyse root system and corresponding protease activity in the rhizosphere of young tomato plants (*Solanum lycopersicum* cv. Moneymaker). The activity of proteases exuded by tomato roots applying in situ zymography was detected along the entire root system. The corresponding root architecture as well as root and shoot biomasses was determined to correlate protease activity with plant growth parameters under varying nitrogen supplies. With higher nitrate fertilisation, the proteases in the rhizosphere were more active than nitrogen-deficient plants. This may indicate that exuded proteases were not solely a plant response to nitrogen deficiency with the aim to increase nitrogen availability. Instead, they may have different roles, e.g. in root development.

Keywords: rhizosphere, protease, nitrate supply, in situ zymography, exudation

1. Introduction

The rhizosphere is most concisely described as the soil influenced by plant roots [1]. It is considered as a dynamic system of interacting processes with major implications for climate and environmental changes in aspects of greenhouse gas emission, carbon sequestration and soil fertility management for sustainable agriculture [2]. Various biotic interactions among plants and microorganisms occur in the rhizosphere, influencing fluxes between organic and inorganic nutrient pools, plant nutrient availability and plant health [3]. Both positive and negative biotic interactions in the rhizosphere are considered to be vitally mediated by root exudates, deeming them a focus in rhizosphere research [4, 5].

Root exudation is a process of excreting substances from plant roots and assumed to be the main source of organic carbon in the rhizosphere [6, 7]. Exudation can occur by rhizodeposition (sloughing off of cells) and passive or active exudation mechanisms of single compounds [8]. Exudation rate and composition have been suggested to vary between root zones with highest rates in apical regions, declining towards older root parts [9]. Other models, however, suggest an even exudation zone surrounding the whole root system [2].

While carbohydrates may constitute the bulk of root exudates, other components like acids, single ions, allochemicals and proteins are of no less importance for rhizosphere processes [4]. In focus of this study are proteases, hydrolytic enzymes cleaving peptide bonds which can be found in all cells and organelles of a plant body [10]. Furthermore, proteases have been shown to occur in the rhizosphere [11]. While bacteria and fungi have long been known to excrete proteases among other extracellular enzymes [12], the knowledge of exudation of proteases by plant roots is more recent [13–15].

It is generally assumed that the main function of both microbial and plant-exuded proteases is nutrient cycling, making nitrogen from organic compounds available for consumption [12, 16]. A second function that has been identified for certain proteases is pathogen defence, working in concert with other lytic enzymes such as lipases and collagenases [15, 17]. A third function still sometimes underestimated is the processing of extracellular proteins, regulating cell growth and development, occurring mostly in the cell wall [18].

Modern techniques allow the detection of enzymatic activity in the rhizosphere, using rhizoboxes [19] and in situ zymography to obtain a potentially realistic impression of enzyme activities, their distribution and intensities within the rhizosphere [11]. This study aimed at localising activities of proteases using gelatin as substrate. It has been shown that plants increase their exudation of corresponding compounds to increase nutrient availability under deficiency [9]. Since exuded proteases are supposed to increase N availability [20], the focus was set on protease activity in the rhizosphere of tomato plants grown under different nitrate regimes.

2. Materials and methods

2.1 Plant material

Tomato seeds (*Solanum lycopersicum* cv. Moneymaker) were germinated and cultivated in sand using nutrient solution [21] with different nitrate regimes (0.5, 1.0, 2.0, 5.0, 7.0, 10.0, 15.0, 20.0 mM $Ca(NO_3)_2$ for plants in pots; 0.0, 2.5 or 10.0 mM $Ca(NO_3)_2$ for plants in rhizoboxes). The nutrient solution without nitrogen contained 3.2 mM $CaSO_4$ and 1 mM $CaCl_2$ to maintain osmolarity. The plants were cultivated in a greenhouse (light/dark rhythm 14–10 h and 28–22°C). Seeds were sown in pairs in small pots and transferred individually either to rhizoboxes after 7 days or to medium pots after 10 days. All plants were watered daily: the plants in rhizoboxes with 50 ml of the corresponding solution and the plants in pots using 200 ml nutrient solution for six plants.

2.2 Rhizoboxes

After 7 days, eight seedlings per nutrient solution (four seedlings only for deionised water) were transferred individually to rhizoboxes made of PTFE (internal dimension 15 × 17 × 1.5 cm, **Figure 1**). The bottom of each rhizobox was perforated to avoid dammed-up water but covered with nylon gauze (pore size 200 μm) to retain sand particles. The removable and transparent front glass pane was mounted by polycarbonate clamps and sealed by O-ring insertion (material NBR70).

For plant cultivation, the rhizobox was filled with sand and moistened with nutrient solution. The pane was removed temporarily to insert the seedling. The pane and the sand surface were covered with pond foil to reduce algal growth. The rhizoboxes were kept inclined by 45° during cultivation with the glass pane pointing downwards (**Figure 1**). Plants were cultivated for another 7 days. This procedure allows the observation of root growth and non-invasive analysis of root exudation.

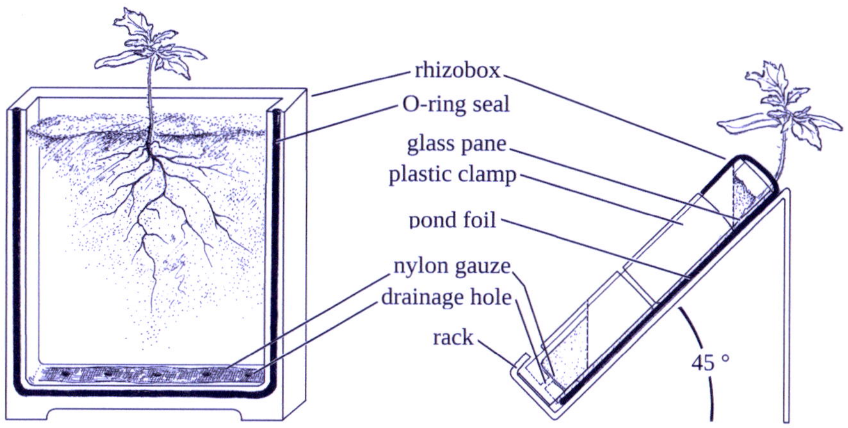

Figure 1.
Design of a rhizobox with removable front lid. Front and side view of a rhizobox with a tomato seedling growing inside the sand-filled box (internal dimensions 15 × 17 × 1.5 cm). The drainage holes at the bottom of the rhizobox were covered by nylon gauze to retain the sand when watering from the top. The front wall was replaced by a glass pane, which was held in place by plastic clamps. An O-ring sealed the gap between the glass and the edges of the rhizobox. The drawing is not to scale.

2.3 In situ soil zymography

2.3.1 Performance of in situ soil zymography

In situ soil zymography was adapted from [11], who used 1% w/v agarose gels with 0.1–0.01% w/v gelatin to determine protease activity. In this study, however, gels with 5% w/v polyacrylamide as matrix and 0.1% gelatin (w/v) as substrate were used. Gelatin was boiled for 10 min to denaturate contaminating enzymes before usage. After polymerisation, gel sizes were adjusted to the inner rhizobox dimensions (16 × 18 × 0.1 cm) and incubated in millipore water for 15 min to remove remaining non-polymerised acrylamide.

Zymographies were performed on four plants per nutrient solution. The rhizoboxes were irrigated with the correspondent solution 2–3 h prior to the experiment. The glass pane was removed carefully, and the root system was documented (Canon PowerShot G7 X) while being illuminated with UV light (365 nm; Blak Ray® B-100 AP, 100 W). Due to the fluorescence of lignified cell walls and phenolic compounds [22], the root system was emphasised (**Figure 2**). Preliminary tests showed no influence on the protease activity due to the irradiation with UV light, as well as the mechanical stress by pushing the glass pane off the root (data not shown).

The gel was placed on top of the root system, locked into position with the pane and wrapped in plastic foil overlaid with a dark cloth. It was incubated for 6 h at 28°C in the growth chamber. After incubation, the pane with the adhering gel was removed, and plants were immediately harvested.

Gels were washed with millipore water for 15 min and stained with 0.1% (w/v) Coomassie brilliant blue R-250 (in 50% (v/v) methanol, 7% (v/v) acetic acid) (modified from [23]) for 15 h at room temperature. Gels were destained (25% (v/v) methanol; 7% (v/v) acetic acid) and watered in deionised water for 15 min before documentation (biostep Felix 2000) on a daylight fluorescent plate. To calculate the remaining gelatin calibration, gels ranging from 0.0 to 0.1% (w/v) gelatin were stained and destained alongside the zymographies.

2.3.2 Calibration of zymographical results

The digital images of the zymography gels and calibration gels were adjusted with GNU Image Manipulation Program (GIMP) 2.8.22 to the same pixel amount per cm using the scale within each image. Image analysation was performed with ImageJ (https://imagej.nih.gov/ij/index.html). All images were converted to 8-bit greyscale.

2.3.3 Non-linear calibration

The average grey value of the calibration gels containing 0 and 0.1% (w/v) gelatin was measured and used as minimum and maximum to adjust the limits of the display range of all images including the zymography gels by using the brightness/contrast tool. Afterwards, the average grey value of each calibration gel slice was calculated using areas of at least 100,000 pixels.

Using the calibration tool of ImageJ, the measured grey values were plotted against the corresponding gelatin concentrations to obtain a calibration curve. This non-linear curve was fit using the logistic regression "Rodbard" and applied to the zymography images. The images were coloured in pseudo-colours for visualisation.

This calibration was preferred for evaluation since the resolution is higher at low–medium gelatin contents of 0–450 ng gelatin (mm^2 gel area)$^{-1}$, while all higher contents merge to form the background noise.

2.3.4 Linear calibration

The ImageJ tool "histogram" was used to analyse and display the number of pixels per grey value within the image. The grey values were then separated into five classes according to the calibration curve. The classes were defined following the intervals of 200 ng gelatin (mm^2 gel area)$^{-1}$ from 0 to 1000 ng gelatin (mm^2 gel area)$^{-1}$ and changed to pseudo-colours accordingly.

This calibration method results in a linear scale of remaining substrate per gel area, yet at a resolution of 200 ng gelatin (mm^2 gel area)$^{-1}$ only. Therefore, the information of nuances amidst the intervals is lost. However, this method may enable a quantitative analysis of soil in situ zymographies in other experimental set-ups and allows a vivid visualisation (**Figure 2**).

2.4 Analysis of zymographies and root systems

Both zymography images and UV photographies were turned to greyscale pictures for the analysis in WinRHIZO (Pro Version; Upgrade 2016a, Regent Instruments Inc.). Each picture contained a length standard for scaling to real values. The parameters obtained from UV photographies were total root length, projected area and the number of root forks per root system. For the zymographies, only the projected area of proteolytic degradation along the root system was obtained for comparison with the root system itself.

2.5 Harvesting of plant biomass

The plants grown in rhizoboxes were harvested immediately after in situ zymography incubation. Plants grown in pots were harvested 25 days after sowing. For all plants, the roots were rinsed thoroughly to remove all sand before shoot, and roots were separated and dried at 70°C for 12 h.

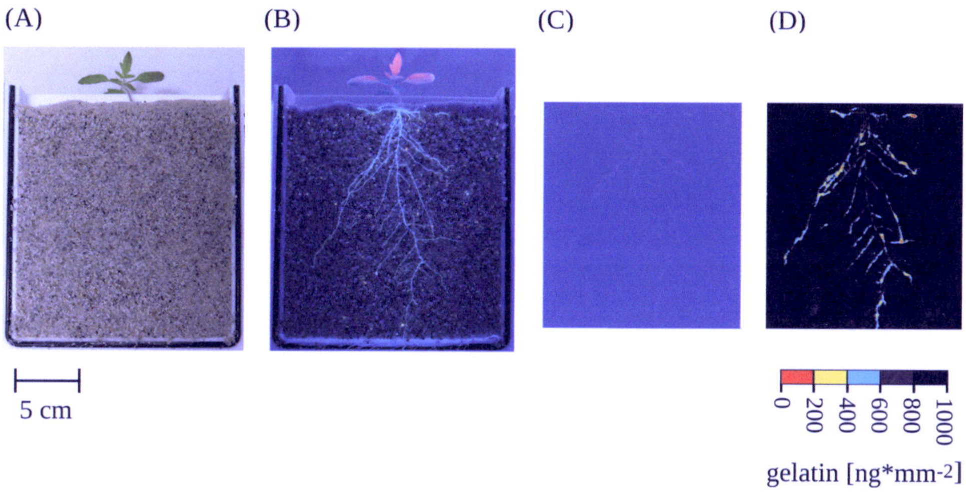

Figure 2.
Tomato root system and corresponding proteolytic activity. (A) Rhizobox rooted by tomato for 7 days and (B) visualisation of the same root system by fluorescence during UV excitation. Corresponding protease activity in the rhizosphere was detected by (C) in situ zymography with gelatin as substrate and Coomassie brilliant blue R-250 staining. (D) The intensity of the proteolytic activity was visualised in pseudo-colours scaled to remaining gelatin [ng mm^{-2}] using a linear calibration.

2.6 Statistical analysis

Experiments were done in randomly blocked design having at least four independent biological samples. ANOVA was performed to reveal statistical significance at 95% confidence level followed by Tukey test for post hoc testing, both using R (Version 3.4.4; 2018, The R Foundation for Statistical Computing) in RStudio (Version 1.1.383; 2009–2017, RStudio Inc.).

3. Results

As in other plants, tomato shoots and roots respond to variation in the nitrogen regime with altered growth and root architecture. To determine the optimum and overload concentration of nitrate for tomato plants, biomass development was analysed after growth in pots and irrigating the plants with nutrient solution containing varying concentrations of $Ca(NO_3)_2$ (**Figure 3**). According to these results, 10 mM nitrate in the nutrient solution was chosen as optimum, and in further experiments, it was more accurately defined as a daily supply of 0.25 mmol nitrate per plant. Excess of nitrate was determined at 20 mM nitrate in the nutrient solution corresponding to 1.0 mmol nitrate per day and plant. Nitrogen deficiency was applied with no additives of nitrate either in nutrient solution or in water. After 7 days cultivation in rhizoboxes (**Figure 2**), nitrogen-deficient plants showed a shoot dry weight that was on average 5 (deionised water) to 8.5 (0 mM nitrate in nutrient solution) times lower than the shoot dry weight of plants with optimum supply (**Figure 4A**). Plants receiving an overload of nitrate showed less shoot dry weight by one third on average than the optimum supply.

The root biomass revealed less difference between the varying cultivation solutions. While the N-deficient plants differed from the optimally supplied plants by a root dry weight 3.7 (deionised water) to 2 (0 mM nitrate in nutrient solution) times less, the root dry weight of plants supplied with a surplus of nitrate did not differ significantly from both optimum- and nitrogen-deficient plants.

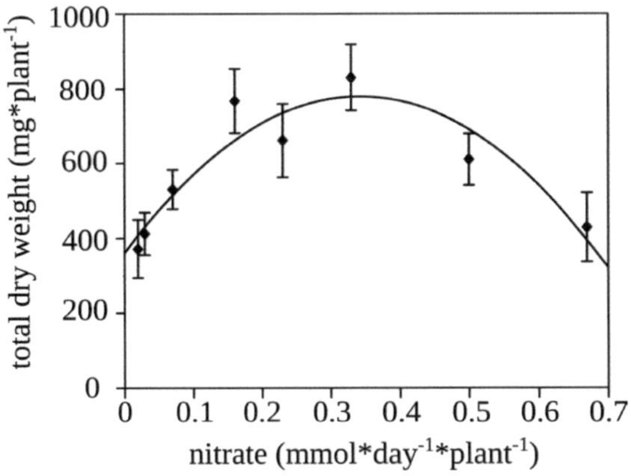

Figure 3.
Effect of nitrate supply on overall dry weight biomass of young tomato plants. The plants were grown in pots and supplied with the correspondent amount of nitrate per day in nutrient solution. The plants were harvested 25 days after sowing. Error bars indicate standard deviations (n = 5 for 0.50 mmol nitrate plant^{-1} day^{-1}, n = 6 for other treatments, $p < 0.001$).

Plants grown under nitrogen deficiency or total deficiency (deionised water) differed significantly in their shoot/root ratio from plants grown under optimum amount and surplus of nitrogen, while no difference was visible within those two groups (**Figure 4B**).

The root architecture of young tomato plants was analysed after 7 days growth in rhizoboxes. With varying supply of nitrogen from deficiency to optimum and overload of nitrate, tomato plants differed highly in root branching (**Figure 5**). When considering the number of root forks per plant (**Figure 5A**), both shortage and surplus of nitrate revealed lower fork counts than optimally supplied plants. While the plants supplied with deionised water did not differ significantly in fork number from the nitrogen-deficient plants, which showed on average three times less forks than plants with optimum nitrate concentration, the plants grown with a surplus of nitrate showed nearly twofold less forks on average than optimally

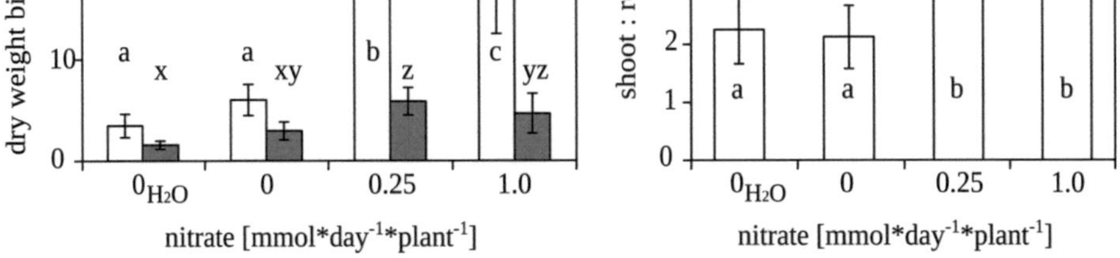

Figure 4.
Effect of nitrate supply on the biomass of young tomato plants. Plants were provided with the correspondent amount of nitrate per day in either deionised water (0_{H_2O}) or nutrient solution and transplanted to rhizoboxes after 1 week of cultivation. The plants were harvested 2 weeks after sowing. (A) Dry weights of shoot (white bars) and root (grey bars) per plant as well as their proportions (B) are shown. Letters indicate group differences at a p-value of <0.001 (n = 4 for deionised water, n = 8 for other nutrient solutions).

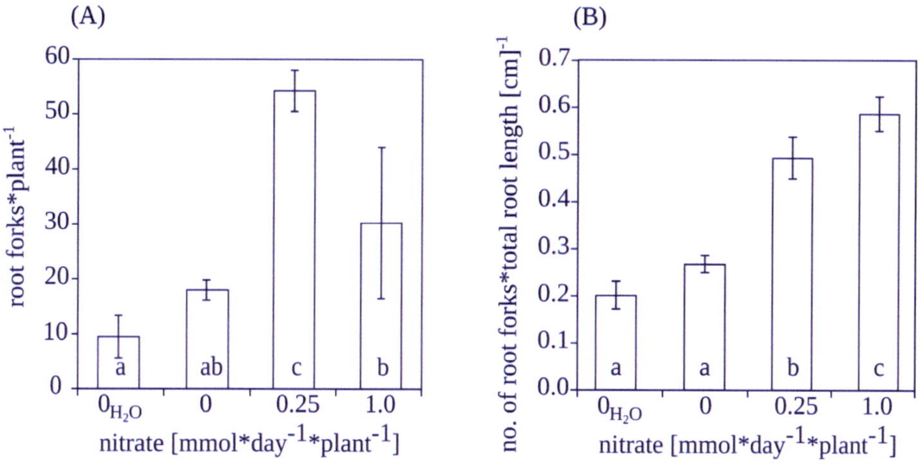

Figure 5.
Effect of nitrate supply on the root architecture of young tomato plants. The plants were supplied with the correspondent amount of nitrate per day in either deionised water (0_{H_2O}) or nutrient solution and transplanted to rhizoboxes after 1 week of cultivation. Harvesting and analyses of the root system were performed after 1 more week. Root lengths were analysed using UV photographies and the software WinRHIZO, and the forks were counted visually. (A) The number of root forks per root system and (B) the number of root forks in relation to root length. Letters indicate group differences at a p-value of <0.001 (n = 4).

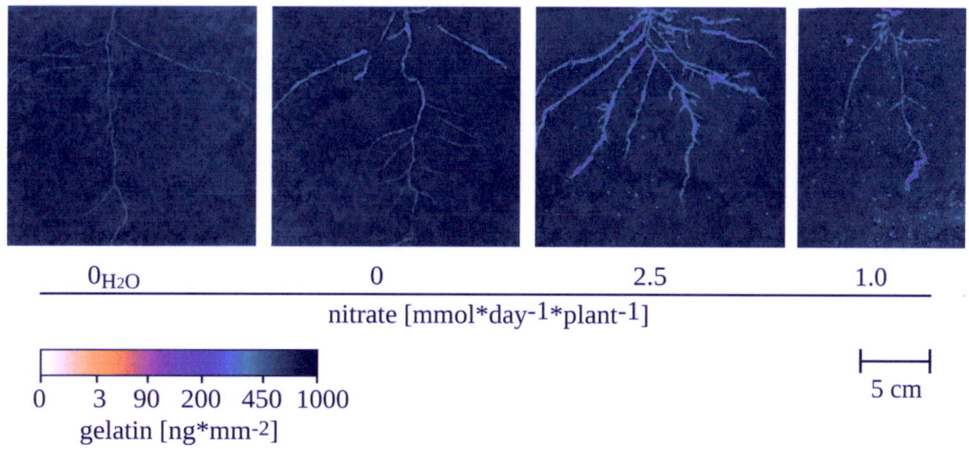

Figure 6.
Effect of nitrate supply during plant growth on protease activity in the rhizosphere of young tomato plants. Exemplary non-linearly calibrated in situ zymographies of root systems of plants grown on deionised water (0_{H_2O}) or different nutrient solutions were converted to pseudo-colours to reveal between different protease activities.

supplied plants. When considering the number of forks in relation to root length, however, plants with both total deficiency and nitrate deficiency showed the lowest number of forks per cm root length, while plants with a surplus of nitrate showed a higher number of forks per cm root length than the optimally supplied plants (**Figure 5B**).

Protease activity in the rhizosphere in relation to plant nitrate supply was visualised along the root system using in situ zymography (**Figure 6**). The area of gelatin degradation was evaluated as proteolytic activity and converted to pseudo-colours to reveal the intensity of proteolytic degradation. This data was used to estimate the area of proteolytic gelatin degradation in relation to the projected root area (**Figure 7**). Plants with nitrogen deficiency (either in deionised water or in nutrient

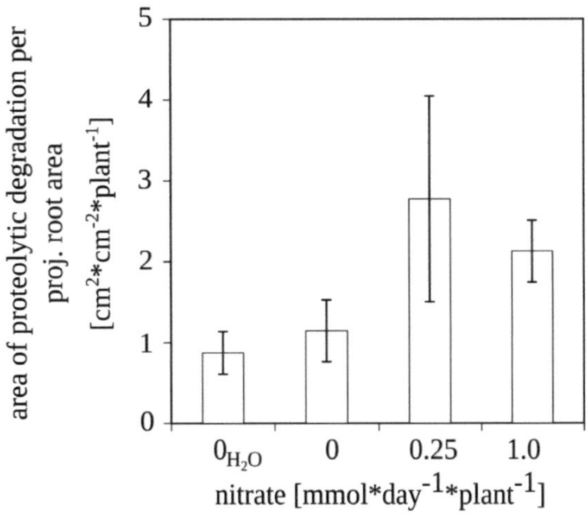

Figure 7.
Effect of plant nitrate supply on protease activity in the rhizosphere of young tomato plants. The comparison of the area of proteolytic activity along the root system to the total projected root area per plant is based on non-calibrated 8bit (greyscale) root zymographies and UV photographies of root systems analysed in WinRHIZO. Letters indicate group differences at a p-value of 0.008 (n = 4).

solution) had an on average 3.2 times lower proteolytic activity per projected root area than well-supplied plants. However, the ratio of proteolytic degradation per root area was not significantly different between optimally fed plants and plants with a surplus of nitrate, clearly indicating that the increase in proteolytic activity was not a response to nitrogen deficiency.

4. Discussion

For understanding of plant health and root adaptation to biotic and abiotic factors, three aspects of the rhizosphere are crucial: root development, biotic interactions and water and nutrient uptake [24]. This study focused on the aspect of nitrogen uptake, inquiring how protease activity in the rhizosphere may depend on nitrate availability. The results obtained using in situ rhizosphere zymography indicate that exuded proteases may serve purposes additional to nitrogen acquisition from organic compounds.

4.1 Plant growth in response to nitrate regime

Plants respond to nitrogen regimes with changes in growth and biomass partitioning as well as in root architecture [25–28], so according parameters were chosen to evaluate the deficient, optimum and excessive nitrogen supply. Plants with nitrogen deficiency showed a strongly reduced shoot biomass compared to optimally supplied plants, while the reduction in root biomass was not as grave (**Figure 4A**). This corresponds well with the general assumption that nitrogen deficiency results in overall reduced biomass due to metabolic limitation. Because nitrogen is an essential element for nucleic acids, proteins and diverse vital molecules from phytohormones to cell wall components like proteoglycans [29], its absence leads to similarity in growth of both total nutrient deficiency (deionised water) and nitrate deficiency only (**Figure 4A**). As an adaptation to nitrogen insufficiency, plants show higher root growth than shoot when nitrogen is low [30]. However, it has been proposed that reduced growth—especially of leaves and shoot—is not only a result of metabolic limitation but an adaptive response to prevent internal starvation [28]. Nitrogen surplus can be detrimental to plants [25, 31] as indicated by reduced shoot biomass of plants with excessive nitrate supply. The root biomass and shoot/root ratio, however, did not differ significantly from optimally supplied plants (**Figure 4A**). Under nitrate overload, total plant growth would be reduced, and a constant C/N ratio has to be maintained [25].

Root architecture

In the applied rhizoboxes, the root system can only expand along the glass pane of the box, resulting in an artificially two-dimensional root system that can be easily evaluated. Changes in the root system architecture (RSA) at different external nitrate concentrations have been shown in *Arabidopsis* [27] but may be present in most plants [28].

The number of root forks was similar for plants supplied with both shortage and excess of nitrate, while plants under optimum nitrate conditions showed higher fork counts as earlier reported for *Arabidopsis* [27]. While low nitrate concentration is assumed to increase root elongation in primary and secondary roots at an overall reduced root weight, high concentrations of nitrate would reduce root elongation at a higher root weight [27], which is consistent with the presented findings. Higher root elongation at comparable root branching under nitrogen deficiency is typical for roots foraging for nitrogen [29] to achieve higher nitrogen uptake efficiency. For high nitrogen supply, on the other hand, the typically reduced branching and growth are assumed to result from nitrogen accumulation in the shoot, inhibiting auxin flux to roots and thus preventing lateral roots to pass an auxin-requiring checkpoint that is vital in lateral root development [32].

4.2 Proteolytic activity in the rhizosphere

While inorganic nitrogen forms such as ammonium and nitrate are usually in focus for plant nitrogen nutrition, plants are also able to take up organic nitrogen in the form of amino acids [28, 33]. Because of root-derived protease activity, even the uptake of small proteins was suggested for root hairs [34]. Plant roots can exude both endo- and exopeptidases [16]. Endopeptidases cleave within peptide chains, while exopeptidases only cleave amino acids at the termini of the substrate [10]. Additionally, proteases can be of different specificity for a substrate, ranging from unspecific to highly regulated hydrolysis, the latter depending on a certain amino acid sequence or pattern [35].

Gelatin as protease substrate is processed mainly by endopeptidases due to the molecule structure. Gelatin is a complex biopolymer made of solubilised collagen, long polypeptide-chains composed of the amino acid triplet gly-x-y with x often being proline and y hydroxyproline [36]. While a total of 16–20 amino acids can occur in gelatin, it is approximately made of 33% glycine and 15–25% proline plus hydroxyproline [37, 38]. Chain lengths of roughly 1000 amino acids [37] hugely increase the probability of endopeptidase activity, while the monotony of the triplet pattern suggests higher rates of unspecific proteolytic activity. The addition of an organic nitrogen source (gelatin) several hours previous to the zymography did not result in any priming effect (data not shown).

The zymographically observed increase in proteolytic activity in the rhizosphere with increasing nitrate availability suggests an additional function of the exuded proteases. Nitrogen acquisition as the main protease function would result in an opposed pattern with high proteolytic activity under nitrate limitation. The controlled growth conditions should also limit pathogen occurrence, hence limiting the number of proteases exuded for pathogen defence. Thus, developmental functions of exuded proteases might be worth investigating. Recently, a subtilase TREXS has been identified in *Nicotiana tabacum* root exudates and is assumed to fulfil a role in root development according to its relative SDD1 in *Arabidopsis* which is involved in stomata development [39]. For the intracellular subtilases XSP1 and AIR1 in *Arabidopsis*, a function in lateral root formation has been proposed due to the specific expression in the respective tissues [40, 41]. Papain-type cysteine endopeptidases are expressed in root epidermis cells that are separated for lateral root emergence. Loss of papain-type endopeptidases AtCEP1 or AtCEP2 in maize caused delayed emergence of lateral root primordia [42].

4.3 Localization of proteolytic activity

The activity detected in the in situ zymographies results from extracellular proteases. Like other extracellular enzymes, extracellular proteases may diffuse from the cell wall of their mother cell to the rhizosphere soil unless they are held back, e.g. by root or microbe mucus [12].

In contrast to other studies [9] reporting high protease activities at root zones of high exudation rates, especially at root tips, the distribution of protease activity was observed roughly along the whole root system in this study. Protease activity was documented for 70–90% of the root length for all treatments. This corresponds with the exudation zone proposed by [2] to surround the whole root system. In grasses, this zone forms the rhizosheath where microbial activity has been suggested to be especially high due to carbon deposits [43]. This phenomenon may also occur in dicots [2], suggesting microbial activity as part of the observed proteolysis in the rhizosphere. However, proteolytic degradation along the root system was also visible with plants grown from surface-sterilised seeds under aseptic conditions (data not shown, plants grew untypically/were deformed). The cultivation of plants on sand instead of microbe-rich soil, the short time span and the greenhouse conditions may all put a limit to microbial colonisation in comparison to natural conditions. Additionally, rhizosphere bacteria usually absorb and assimilate nitrate only in the absence of either organic N—e.g. plant exudates—or ammonium [44], while it is a major N source for most higher plants [45]. Nitrate might even decrease microbial growth in comparison to ammonium or anorganic nitrogen compounds [46, 47]. Hence, the participation of bacteria in proteolytic activity in rhizobox experiments can be assumed to be of limited importance.

Since the root system was always clearly outlined in the zymographies, the spatial definition for the rhizosphere observed in this context is straightforward and restricted to those areas of proteolytic degradation that were linked to the root system. Apart from this, however, minor to moderate proteolytic degradation could also be observed at small, random spots across the substrate. These might coincide with colonies of microorganisms. Interestingly, the intensity of the proteolytic degradation occurring in these spots increased with increasing nitrogen supply, suggesting higher growth rates at higher nitrogen supply. In comparison to proteolytic activity in the rhizosphere, however, both size and number of these spots were very low.

5. Conclusion

Differences in protease activity between the nitrogen treatments could be a result of two different regulative possibilities: firstly, the differences in protease amounts—based on expression and exudation of the proteases—and, secondly, the differences in activity of the available proteases depending on abiotic factors and protease processing [48]. Since proteolytic activity in the rhizosphere of young tomato plants increased with increasing nitrate availability, a function of the observed proteases for nitrogen acquisition seems unlikely. The supplication of nitrogen as inorganic nitrate alone probably limited the microbial growth in the rhizosphere and additionally avoided any priming for organic N hydrolysis.

Author details

Hannah Holzgreve, Manuela Eick and Christine Stöhr*
University of Greifswald, Greifswald, Germany

*Address all correspondence to: stoehr@uni-greifswald.de

References

[1] Hiltner L. Über neuere Erfahrungen und Probleme auf dem Gebiete der Bodenbakteriologie unter besonderer Berücksichtigung der Gründüngung und Brache. Arbeiten der Deutschen Landwirtschafts-Gesellschaft. 1904;**98**:59-78

[2] York LM, Carminati A, Mooney SJ, Ritz K, Bennett MJ. The holistic rhizosphere: Integrating zones, processes, and semantics in the soil influenced by roots. Journal of Experimental Botany. 2016;**67**(12):3629-3643

[3] Bakker MG, Schlatter DC, Otto-Hanson L, Kinkel LL. Diffuse symbioses: Roles of plant–plant, plant–microbe and microbe–microbe interactions in structuring the soil microbiome. Molecular Ecology. 2014;**23**:1571-1583

[4] Badri DV, Vivanco JM. Regulation and function of root exudates. Plant, Cell and Environment. 2009;**32**:666-681

[5] Sasse J, Martinoia E, Northen T. Feed your friends: Do plant exudates shape the root microbiome? Trends in Plant Science. 2018;**23**(1):25-41

[6] Hutsch BW, Augustin J, Merbach W. Plant rhizodeposition an important source for carbon turnover in soils. Journal of Plant Nutrition and Soil Science. 2000;**165**:397-407

[7] Nguyen C. Rhizodeposition of organic C by plants: Mechanisms and controls. Agronomie. 2003;**23**:375-396

[8] Baetz U, Martinoia E. Root exudates: The hidden part of plant defense. Trends in Plant Science. 2014;**19**(2):90-98

[9] Egamberdieva D, Renella G, Wirth S, Islam R. Enzyme activities in the rhizosphere of plants. In: Shukla G, Varma A, editors. Soil Enzymology. Berlin Heidelberg: Springer; 2011. pp. 149-166

[10] van der Hoorn R. Plant proteases: From phenotypes to molecular mechanisms. Annual Review of Plant Biology. 2008;**59**:191-223

[11] Spohn M, Carminati A, Kuzyakov Y. Soil zymography—A novel *in situ* method for mapping distribution of enzyme activity in soil. Soil Biology and Biochemistry. 2013;**58**:275-280

[12] Burns RG, DeForest JL, Marxsen J, Sinsabaugh RL, Stromberger ME, Wallenstein M, et al. Soil enzymes in a changing environment: Current knowledge and future directions. Soil Biology and Biochemistry. 2013;**58**:216e234

[13] Godlewski M, Adamczyk B. The ability of plants to secrete proteases by roots. Plant Physiology and Biochemistry. 2007;**45**:657-664

[14] Song Y, Ling N, Ma J, Wang J, Zhu C, Raza W, et al. Grafting resulted in a distinct proteomic profile of watermelon root exudates relative to the un-grafted watermelon and the rootstock plant. Journal of Plant Growth Regulation. 2016;**35**(3):778-791

[15] Guo B, Wang H, Yang B, Jiang W, Jing M, Li H, et al. *Phytophthora sojae* effector PsAvh240 inhibits host aspartic protease secretion to promote infection. Molecular Plant. 2019;**12**(4):552-564

[16] Adamczyk B, Godlewski M, Smolander A, Kitunen V. Degradation of proteins by enzymes exuded by *Allium porrum* roots—A potentially important strategy for acquiring organic nitrogen by plants. Plant Physiology and Biochemistry. 2009;**47**:919-925

[17] Karimi Jashni M, Mehrabi R, Collemare J, Mesarich CH, de Wit PJGM. The battle in the apoplast: Further insights into the roles of proteases and their inhibitors in plant–pathogen interactions. Frontiers in Plant Science. 2015;**6**:584

[18] Flinn BS. Plant extracellular matrix metalloproteinases. Functional Plant Biology. 2008;**35**:1183-1193

[19] Boukcim H, Pages L, Plassard C, Mousain D. Root system architecture and receptivity to mycorrhizal infection in seedlings of *Cedrus atlantica* as affected by nitrogen source and concentration. Tree Physiology. 2001;**21**:109-115

[20] Adamczyk B, Smolander A, Kitunen V, Godlewski M. Proteins as nitrogen source for plants. Plant Signaling & Behavior. 2010;**5**(7):817-819

[21] Stöhr C, Ullrich WR. A succinate-oxidising nitrate reductase is located at the plasma membrane of plant roots. Planta. 1997;**203**:129-132

[22] Buschmann C. Imaging of the blue, green, and red fluorescence emission of plants: An overview. Photosynthetica. 2000;**38**(4):483-491

[23] Troeberg L, Nagase H. Zymography of metalloproteinases. Current Protocols in Protein Science. 2003;**33**(1):21.15.1-21.15.12

[24] Hodge A, Berta G, Doussan C, Merchan F, Crespi M. Plant root growth, architecture and function. Plant and Soil. 2009;**321**(1-2):153-187

[25] Stöhr C. Relationship of nitrate supply with growth rate, plasma membrane-bound and cytosolic nitrate reductase, and tissue nitrate content in tobacco plants. Plant, Cell and Environment. 1999;**22**:169-177

[26] Nagel OW, Konings H, Lambers H. Growth rate and biomass partitioning of wildtype and low-gibberellin tomato (*Solanum lycopersicum*) plants growing at a high and low nitrogen supply. Physiologia Plantarum. 2001;**111**:33-39

[27] Linkohr BI, Williamson LC, Fitter AH, Leyser HMO. Nitrate and phosphate availability and distribution have different effects on root system architecture of *Arabidopsis*. The Plant Journal. 2002;**29**(6):751-760

[28] Nacry P, Bouguyon E, Gojon A. Nitrogen acquisition by roots: Physiological and developmental mechanisms ensuring plant adaptation to a fluctuating resource. Plant and Soil. 2013;**370**(1):1-29

[29] Krapp A. Plant nitrogen assimilation and its regulation: A complex puzzle with missing pieces. Current Opinion in Plant Biology. 2015;**25**:115-122

[30] Reynolds HL, D'Antonio C. The ecological significance of plasticity in root weight ratio in response to nitrogen. Plant and Soil. 1996;**185**:75-97

[31] Elia A, Conversa G. Agronomic and physiological responses of a tomato crop to nitrogen input. European Journal of Agronomy. 2012;**40**:64-74

[32] Forde BG. Nitrogen signalling pathways shaping root system architecture: An update. Current Opinion in Plant Biology. 2014;**21**:30-36

[33] Näsholm T, Kielland K, Ganeteg U. Uptake of organic nitrogen by plants. Tansley Review. New Phytologist. 2009;**182**:31-48

[34] Paungfoo-Lonhienne C, Lonhienne TGA, Rentsch D, Robinson N, Christie M, Webb RI, et al. Plants can use protein as a nitrogen source without assistance from other organisms. Proceedings of the National Academy of Sciences of the United States of America. 2008;**105**(11):4524-4529

[35] Song J, Tan H, Perry AJ, Akutsu T, Webb GI, Whisstock JC, et al. PROSPER: An integrated feature-based tool for predicting protease substrate cleavage sites. PLoS One. 2012;**7**(11):e50300

[36] Gómez-Guillén MC, Giménez B, López-Caballero ME, Montero MP. Functional and bioactive properties of collagen and gelatin from alternative sources: A review. Food Hydrocolloids. 2011;**25**:1813-1827

[37] Duconseille A, Traikia M, Lagrée M, Jousse C, Pagès G, Gatellier P, et al. The impact of processing and aging on the oxidative potential, molecular structure and dissolution of gelatin. Food Hydrocolloids. 2017;**66**:246-258

[38] Farris S, Song J, Huang Q. Alternative reaction mechanism for the cross-linking of gelatin with glutaraldehyde. Journal of Agricultural and Food Chemistry. 2010;**58**(2):998-1003

[39] Wendlandt T, Moche M, Becher D, Stöhr C. A SDD1-like subtilase is exuded by tobacco roots. Functional Plant Biology. 2016;**43**(2):141-150

[40] Neuteboom LW, Ng JMY, Kuyper M, Clijdesdale OR, Hooykaas PJJ, van der Zaal BJ. Isolation and characterization of cDNA clones corresponding with mRNAs that accumulate during auxin-induced lateral root formation. Plant Molecular Biology. 1999;**39**:273-287

[41] Zhao C, Johnson BJ, Kositsup B, Beers EP. Exploiting secondary growth in *Arabidopsis*. Construction of xylem and bark cDNA libraries and cloning of three xylem endopeptidases. Plant Physiology. 2000;**123**:1185-1196

[42] Höwing T, Dann M, Mueller B, Helm M, Scholz S, Schneitz K, et al. The role of KDEL-tailed cysteine endopeptidases of Arabidopsis (AtCEP2 and AtCEP1) in root development. PLoS One. 2018;**13**(12):e0209407

[43] Luque-Almagro VM, Gates AJ, Moreno-Vivián C, Ferguson SJ, Richardson DJ, Roldán MD. Bacterial nitrate assimilation: Gene distribution and regulation. Biochemical Society Transactions. 2011;**39**(6):1838-1843

[44] Bloom AJ. The increasing importance of distinguishing among plant nitrogen sources. Current Opinion in Plant Biology. 2015;**25**:10-16

[45] Mahmood T, Kaiser WM, Ali R, Ashraf M, Gulnaz A, Iqbal Z. Ammonium versus nitrate nutrition of plants stimulates microbial activity in the rhizosphere. Plant and Soil. 2005;**277**(1-2):233-243

[46] Ai C, Liang G, Sun J, Wang X, Zhou W. Responses of extracellular enzyme activities and microbial community in both the rhizosphere and bulk soil to long-term fertilization practices in a fluvo-aquic soil. Geoderma. 2012;**173-174**:330-338

[47] Denef K, Roobroeck D, Manimel Wadu M, Lootens P, Boeckx P. Microbial community composition and rhizodeposit-carbon assimilation in differently managed temperate grassland soils. Soil Biology and Biochemistry. 2009;**41**:144-153

[48] Nannipieri P. Role of stabilised enzymes in microbial ecology and enzyme extraction from soil with potential applications in soil proteomics. In: Nannipieri P, Smalla K, editors. Nucleic Acids and Proteins in Soil, Soil Biology. Vol. 8. Berlin, Heidelberg: Springer-Verlag; 2006. pp. 75-94

Chapter 2

Morphological and Physiological Root Plasticity and Its Relationships with Shoot Growth of Rice with Water Regimes and Microbial Densities

Abha Mishra

Abstract

There is renewed interest in root research for undergirding a second Green Revolution. The modular nature of root systems makes them amenable to both morphological and/or physiological plasticity when encountering heterogeneous environments. Such plasticity, the ability to change and adapt in response to variations in the underground environment, is linked to a shoot response and to consequent dry matter production, which is an important subject of research. This exploration is relevant in paddy production, especially in the context of climate change where rice production needs to be intensified with reduced water application and with reduced methane emission. This chapter reviews the plastic response of roots and illustrates some preliminary findings on the effects of biotic (soil microbes) and abiotic (water regimes) stimuli on root growth and activity and their relationships with shoot growth and its implications for mitigation of methane production without compromising grain yield.

Keywords: root plasticity, paddy, rice physiology, methane emission, climate change

1. Introduction

There is renewed interest in root research for undergirding a second Green Revolution. An article in *Nature* [1] reports on four of the most promising ways for boosting food production through modifications in roots: designer roots, stealth scavengers, microbial manipulation, and healthy fixation. All four ways, involving genetic manipulation of belowground traits, are undergoing evaluation. However, environmentally induced phenotypic variation in plants is often observed and is considered to be a functional response that maximizes fitness in variable environments. Such response is termed as reaction norms. The reaction norms may or may not be plastic. If it is plastic, then plasticity and reaction norms are used interchangeably. Basically, it refers to the set of phenotypes that can be produced by an individual genotype that is exposed to different environmental conditions [2].

Various studies have shown the modular nature of root systems that make them amenable to both morphological and/or physiological plasticity when encountering heterogeneous environments. For example, a variety of crops proliferate roots in areas of high nutrient concentration, and that plant nutrient concentration and yield could be higher in heterogeneous soil than in homogenous soil [3, 4]. Increased uptake and growth responses were attributed both to root proliferation increasing uptake potential and to the fact that a given soil volume has the limited binding capacity and thus as nutrient supply increases [5]. Such root proliferation in response to locally elevated soil resource levels is simply one example of morphological plasticity to an environmental signal, one of many forms of phenotypic plasticity exhibited by plants. Such malleability, the ability to change and adapt in response to variations in the underground environment, means that many root traits are tailored by their environment.

With increasing evidence that environmental heterogeneity is increasing, due to climate change effects, it is important to investigate how roots will respond in different environments. In particular, root traits linked to shoot growth and dry matter production should be investigated in order to understand how the roots' plasticity can have a role in enhancing grain yield in a dynamic soil environment and whether roots' morphological and physiological plasticity is linked to a shoot response and to consequent dry matter production. This is an important and increasingly relevant subject of research especially in paddy cultivation where there is an increasing demand to grow rice with reduced water application.

With regard to dry matter production in rice, there are several reports that show that many shoot morphological and physiological traits contribute to high yield, such as larger sink size, higher leaf area index, larger leaf area duration, higher photosynthetic rate, slower leaf senescence, stronger lodging resistance, greater biomass accumulation before heading, and more translocation of carbohydrates from the vegetative parts to the panicle during the grain-filling period. Fewer studies have been conducted on root morphological and physiological traits that may be linked with shoot growth and yield. Studies so far conducted mainly consider the effects of genetic variability on root morphological and physiological traits [6].

These studies do not reflect the effects of the soil environment that could greatly alter root architecture since root systems' growth and functioning are regulated not only by genetic programs but also are influenced by abiotic and biotic stimuli [7]. In particular, root traits that are linked to shoot growth and dry matter production should be investigated in order to understand the roots' plasticity and their possible role in enhancing grain yield in a dynamic soil environment.

Given the climate variability, methane emission from paddy fields, and water constraints facing the rice sector in many countries, the most important crop management practice which has got major attention, both from farmers and researchers, is the cessation of continuous flooding, either through intermittent irrigation or by keeping soil moist but not continuously inundated. The intermittent irrigation of rice is not something new, and recently it has been supported in some rice-producing countries in an attempt to reduce the volume of irrigation water used [8].

Some earlier reports on the effects of cessation of flooding have suggested that under unsaturated soil moisture conditions, there is a significant decrease in dry matter production and grain yield for rice [9, 10]. It is suggested that this could be due to a rapid rate of loss of nitrogen facilitated by nitrification and denitrification [11]. However, others have reported a higher yield correlated with intermittent irrigation during the vegetative stage when accompanied by SRI management practices (transplanting younger seedlings 1–2/hill, avoiding continuous soil saturation, aerating soil, and applying organic manure as much as possible) due to healthy root growth and greater soil microbial activity [12, 13] and even under post-anthesis water-deficit conditions when organic matter has been applied to the soil [14]. It has been reported that skillful soil drying post-anthesis improves remobilization of carbon reserve and grain filling [15].

Such inconsistent reports on the effects of intermittent irrigation and/or nonflooded water regimes for rice production leave some important questions unanswered since they did not assess how rice plants' roots and shoots will respond, respectively and jointly, when subjected to different soil moisture conditions in combination with varying soil microbial condition. There is limited information whether these responses, if they occur, will lead to greater dry matter production or to less and whether these recommended practices and resultant morphological and physiological plasticity can have any contribution toward mitigation of methane emission from the rice fields. This is a research area warranting investigation.

This chapter reports some initial research findings on the plastic response of rice plants that resulted due to change in water regimes and microbial density. Further, it illustrates the causal relationship between rice root and shoot growth and also discusses the implication of root plasticity for mitigation of methane emission from rice fields. The study was conducted to assess the effects of differences in the soil biota in conjunction with alternative water management practices in rice.

In this context, the term alternative water management practices have been introduced here as "water-saving irrigation" to describe producing more rice with less water. This involves (i) reducing the depth of ponded water; (ii) keeping the soil just saturated, not continuously flooded; or (iii) employing intermittent irrigation or alternate wetting and drying, i.e., allowing the soil to dry out to a certain extent before reapplying irrigation water.

2. Methodology

Black clay soil was collected from the rice research farm of the Asian Institute of Technology where the previous crop grown was rice. The average composition of the soil was 10.2% sand, 23.2% silt, and 66.2% clay, with pH (1.1) of 5.0. Organic C was 1.38%, total N 0.14%, available P 11 mg kg^{-1}, and available K 212 mg kg^{-1}. Cation exchange capacity was 22.6 cmol kg^{-1}.

After air drying, the soil samples were crushed, and crop residues were removed by hand. In each plastic pot (60 cm high with diameters 50 cm at the top and 40 cm at the bottom), 65 kg of soil was placed. All pots were flooded by the addition of distilled water to a depth of 3–4 cm for a week before transplanting and were dressed with 138 mg *N* and 12.3 mg *P* per kg of soil applied in NPK fertilizer 16:16:0 as basal application and urea (46:0:0) at 15 and 45 days after transplanting (DAT). Single 15-day-old seedlings (variety *Pathumthani*: maturity period = 120 days, nonphotosensitive) with two fully expanded leaves grown in a dry seedbed were transplanted within 2 hours of uprooting from the nursery seedbed with a sowing depth of 1.5 cm. Water treatment was started 7 days after transplanting when transplanting shock had disappeared.

3. Experiment 1

The first experiment was set up to evaluate the effect of water regimes. Root length density, root-oxidizing activity rate, and chlorophyll content of lower leaf were studied under four water regimes:

1. Intermittent flooding (IF-I)—Pots were maintained with 5 cm depth of water from the soil surface and maintained for 12 days, then drained for 3 days, and

again reflooded with the same depth of ponded water. Three 3-day drying periods were provided at 19, 34, and 50 days after transplanting (DAT) followed by flooded water treatment (5 cm water depth continuously) until maturity.

2. Intermittent flooding (IF-II)—In another pot, similar procedure like IF-I was followed for 5 times at 19, 34, 50, 66, and 82 DAT, followed by flooded water until maturity.

3. Nonflooded (NF)—Pots were maintained under the continuous nonflooded condition and at field capacity (FCp) at the rooting zone.

4. Continuous flooded (CF)—5 cm depth of ponded water was maintained until maturity.

For root study, soil samples were collected from pots at flowering (72 DAT) and at 20 days after flowering (DAF), i.e., at 92 DAT, from the upper (15–20 cm) and subsoil (35–40 cm) layers for root length and root-oxidizing activity.

Roots after being washed with water were cut into small pieces. The root length was calculated using the line intersection method described by Tennant [16]. Root length density (RLD) was then calculated by using the formula: RLD = RL/V, where RL = root length and V = volume of the soil core soil.

Root activity (ROA) rate was measured by assaying the oxidation of alpha-naphthylamine. Five grams of fresh roots were transferred into a 150 ml flask containing 100 ml of 20 mg l^{-1} alpha-naphthylamine. The flask was incubated for 4 hours at room temperature (25 ± 1°C) in an end-over-end shaker. After incubation, the aliquots were filtered, and 2 ml of aliquots was reacted with 10 ml of 0.1% sulfanilic acid and then with 2 ml of 50 ppm $NaNO_3$. The resultant color was measured by spectrophotometer at 530 nm, and the value is expressed as μg $(g\ Fw)^{-1}\ h^{-1}$.

Chlorophyll content of the flag leaf and of the third leaf was recorded at intervals of 7 days from flowering to physiological maturity stages, using a chlorophyll meter (SPAD 502; Minolta Corp; Tokyo) calibrated by using spectrophotometric assays in order to determine the exponential equation to directly convert its output to leaf chlorophyll concentration [16]. These data were collected from undisturbed pots for each treatment combination which had not been used for root study and nitrogen estimation.

4. Experiment 2

In another experiment, IF-I, IF-II, and CF water regimes were tested with three soil conditions that differed in soil microbial density. The three soil conditions were untreated normal soil (NS), autoclaved soil (AUS) in which soil biota had been mostly minimized, and soil in which the abundance of soil biota had been enhanced by applying a solution of effective microorganisms (EMS).

A commercial preparation of effective microorganisms known as "Bio EM" was obtained from EMRO Thailand. The Bio EM was prepared by using a concentrated stock solution of effective microorganisms, EM-1. The formulation of EM-1 is kept secret, although according to one of the EMRO centers (BIONOVA Hygiene GmbH, Stans, Switzerland), EM-1 contains 1.3×10^7 colony-forming units (cfu) of lactic acid bacteria ml^{-1}, 3.3×10^4 cfu photosynthetic bacteria ml^{-1}, and 1.3×10^4 cfu of yeast ml^{-1}. Bio EM was processed from EM-1 by fermentation under anaerobic conditions with water and sugarcane molasses for 7 days.

In the EMS soil pots, the Bio EM solution was first applied at 7 DAT, with 6.75 ml of concentrated EM solution mixed in 4.5 l of water. Before the start of any irrigation of these trials with EM-treated soil (EMS), 0.5 l of this mixed solution was applied. After that, water levels were maintained in all EMS pots according to the treatment schedules. The EM application was repeated at weekly intervals until 1 month before harvesting, unless a draining period coincided with the EM application. EM application was avoided during draining periods and was applied with the next scheduled irrigation, immediately following a drainage period.

For further experimental details, see [17].

5. Results

5.1 Effects of varying water regimes

The result indicated significant effects of varying water regimes on root length density, both at the upper and subsoil layer. At flowering, there was no difference recorded in the root distribution in the upper soil depth in the intermittent irrigation followed until vegetative stage (IF-I) and continuously flooded treatments (CF), and in these treatments, most of the roots were observed to be distributed in the upper soil layer. In contrast, fewer roots were observed at lower soil depth in the CF than the IF-II. The distribution pattern was different in the nonflooded treatment (NF) treatments compared to the other three water regimes. In this treatment, almost half of the total root length density was distributed at the lower soil depth. At the later growth stage, a drastic reduction in root growth was observed under the continuously flooded treatment compared to other water regimes at both soil depths. Almost 70% root reduction was observed under the continuously flooded condition in the upper soil depth (**Figure 1**).

Further, it was observed that the physiological activity of the roots, i.e., root-oxidizing activity rate, was higher in the IF-I water regime than in the continuously flooded condition and continuously intermittent irrigation at a later growth stage (**Figure 2**). The experiment revealed that there was a positive correlation between chlorophyll content of lower leaves and root activity in all water regimes (**Figure 3**) depicting the causal relationship with those shoot traits which are linked to increased dry matter production.

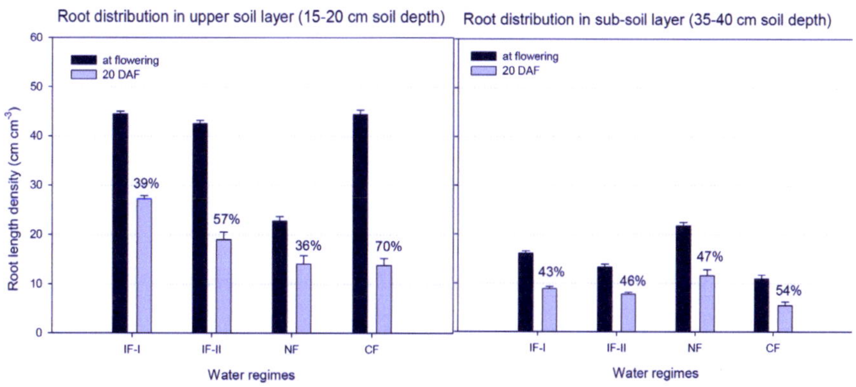

Figure 1.
Root length density (RLD) (cm/cm^3) in the upper and subsoil layer at flowering and 20 days after flowering of rice plant grown in pots under different water regimes (IF-I, IF-II, NF, and CF). The number above the gray bars shows percentage reduction in root length density at 20 days after flowering. Error bars show SE.

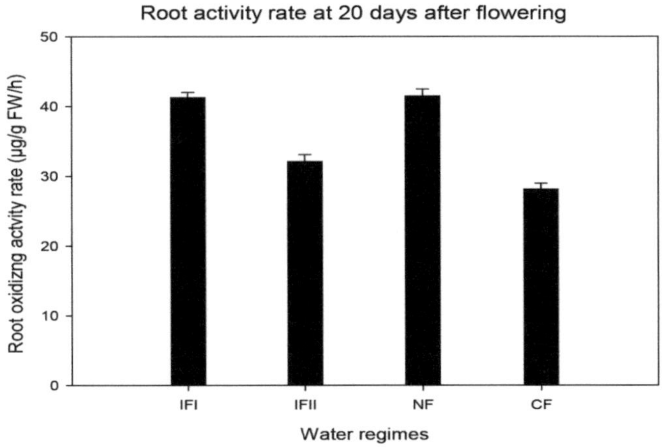

Figure 2.
Root-oxidizing activity rate under varying water regime. IF-I, intermittent draining three times; IF-II, intermittent draining five times; NF, nonflooded; and CF, continuous flooding.

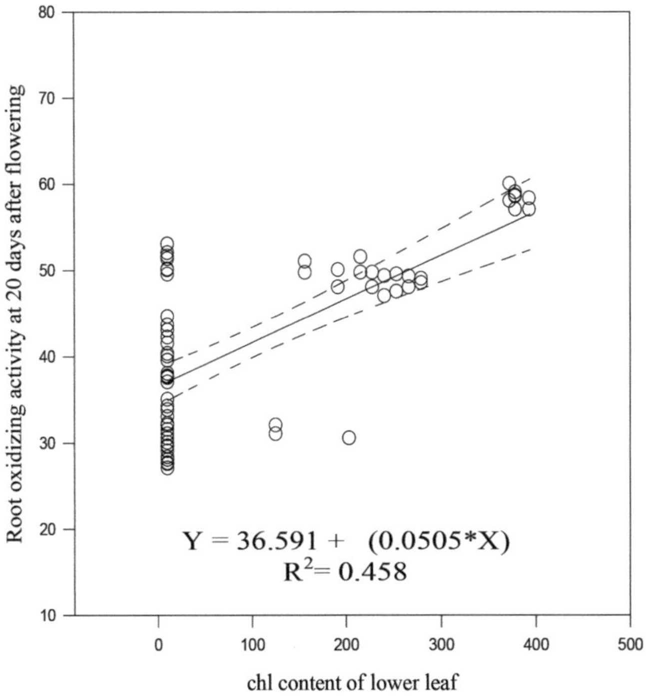

Figure 3.
Relationship between chlorophyll content of lower leaf and root-oxidizing activity at 20 days after flowering.

5.2 Effects of varying water regimes and soil microbial density

EMS treatment increased the number of spikelets/panicle and filled grains/panicle under all water regimes. Also, at the flowering stage, both root length density and root activity were higher in EM-treated rice plants under all three water regimes evaluated. However, at later growth stages, the EM-treated plants grown under IF-I and IF-II showed lower root activity rates compared to plants that were grown in autoclaved or normal soil (**Table 1**).

Water regimes	**Soil conditions**		
	Normal soil	**EM-treated soil**	**Autoclaved soil**
Number of spikelets per panicle			
IF-I	227.75 ± 2.09	244.00 ± 1.90	203.63 ± 1.82
IF-II	187.38 ± 1.16	196.13 ± 1.72	166.50 ± 1.22
NF	191.50 ± 1.27	206.25 ± 1.74	173.88 ± 2.36
Filled grains per panicle			
IF-I	227.75 ± 2.09	244.00 ± 1.90	203.63 ± 1.82
IF-II	187.38 ± 1.16	196.13 ± 1.72	166.50 ± 1.22
NF	191.50 ± 1.27	206.25 ± 1.74	173.88 ± 2.36
RLD at 15–20 cm soil depth (at flowering)			
IF-I	45.46 ± 0.78	63.51 ± 1.75	54.83 ± 1.36
IF-II	45.53 ± 1.59	62.06 ± 1.26	53.38 ± 1.21
NF	22.77 ± 1.61	23.57 ± 1.70	22.18 ± 2.03
Root-oxidizing activity rate (μg/g FW/h)at flowering			
IF-I	63.40 ± 0.74	66.31 ± 0.77	64.35 ± 0.93
IF-II	63.10 ± 0.35	66.53 ± 0.38	64.20 ± 0.40
NF	53.50 ± 0.49	59.23 ± 0.43	52.31 ± 0.56
Root-oxidizing activity rate (μg/g FW/h)at 20 days after flowering			
IF-I	42.28 ± 0.57	40.01 ± 1.49	58.30 ± 0.35
IF-II	33.1 ± 0.33	32.10 ± 0.58	33.54 ± 0.46
NF	41.31 ± 0.57	40.15 ± 0.60	49.70 ± 0.44
Grain weight per plant (gm/pot)			
IF-I	165.74 ± 4.07	186.23 ± 5.2	198.27 ± 4.37
IF-II	101.75 ± 7.28	115.41 ± 7.11	103.37 ± 6.11
NF	116.46 ± 2.19	134.59 ± 5.6	125.71 ± 1.58
Total biomass (g/plant)			
IF-I	341.71 ± 10.22	363.67 ± 15.57	398.9 ± 20.76
IF-II	215.73 ± 7.67	201.38 ± 9.37	218.61 ± 4.59
NF	248.14 ± 9.75	257.62 ± 5.29	243.95 ± 11.13

IF-I, intermittent draining three times; IF-II, intermittent draining five times; and NF, nonflooded, and varying soil conditions: NS, normal soil; EMS, soil treated with effective microorganisms solution; and AUS, autoclaved soil on morphological and physiological root, shoot traits, and grain yield. Values show mean ± SE.

Table 1.
Effect of varying water regimes and soil types on morphological and physiological root, shoot traits and grain yield.

To gain further understating, soil nitrogen status was studied. At flowering, an appreciable increase in the concentration of available nitrogen (*N*), and of NH_4^+ in particular, was found in the rhizosphere soil with EMS treatment under IF-I and IF-II water regimes, but not with NF (**Table 2**). This indicates the possible impact of drying and rewetting of soil on the microbial populations. It seems that repeated application of EM solution in the EMS pots increased the amount of soil-available *N* due to the rapid rate of mineralization. However, at 20 days after flowering, the concentration of *N* was higher in the AUS treatment compared to EMS and NS

	Intermittent draining, three times (IF-I)		Intermittent draining, 5 times (IF-II)		Nonflooded (NF)	
	F	20 DAF	F	20 DAF	F	20 DAF
Autoclaved soil (AUS)	48.76 (64.62)	55.3	37.09 (47.19)	13.72	76.77 (39.71)	48.48
EM-treated soil (EMS)	37.21 (91.09)	23.38	25.23 (64.87)	8.1	66.64 (42.6)	24.24
Normal soil (NS)	56.32 (56.31)	43.62	41.1 (42.6)	10.66	74.71 (37.33)	36.04

IF-I, intermittent draining three times; IF-II, intermittent draining five times; and NF, nonflooded, and soil types: NS, normal soil; EMS, soil treated with effective microorganisms solution (EM); and AUS, autoclaved soil on soil nitrogen status (NH_4^+ and NO_3^-) – N (mg kg^{-1}) at flowering (F) and total available nitrogen (mg kg^{-1}) at 20 days after flowering (DAF). The number under parenthesis shows the content of NH_4^+ at flowering.

Table 2.
Effect of varying water regimes and soil types on soil nitrogen status.

under IF-I and NF water regimes. Significant differences at both growth stages were observed in the IF-I and NF water regimes, but not in the IF-II regime, probably due to the higher rate of nitrogen loss from the soil, facilitated by a greater number of repeated drying-wetting cycles.

Within IF-I, the low availability of *N* in the EMS soil at 20 days after flowering indicated that either (a) the plants' *N* uptake rate was higher in EMS soil compared to AUS and NS soil, or (b) competition between plant roots and soil microbes was increased for the *N* at later growth stages due to higher microbial population and thus to a higher rate of immobilization of NH_4^+, or (c) the rate of denitrification was increased after reflooding due to reduced soil conditions and a relatively higher rate of oxygen demand by microorganisms. Upon flooding, under the reduced soil conditions of IF-I, the possibility of leaching loss of NO_3^- was very small. Hence, it seems that either this *N* was taken up by the plant or any remaining nitrate moving downward after reflooding could have been intercepted by anaerobic microorganisms to use as terminal electron acceptors for anaerobic respiration.

In the IF-II water regime, soil nitrogen content was lower than either IF-I or NF water regimes, probably due to an increased rate of nitrogen loss caused by the greater number of times that draining-reflooding was done, facilitating a higher rate of denitrification and immobilization compared to the IF-I and NF water regimes.

Assessment of the available forms of *N* in AUS and NS soils under IF-I and IF-II water regimes at flowering stage indicated that almost half of the nitrogen was present in NO_3^- form, whereas with EMS, the percentage of NH_4^+ was greater (**Table 2**). The higher percentage of NH_4^+ in the EMS treatments reflects a higher rate of mineralization due to higher soil microbial populations.

The presence of significant amounts of NO_3^- in the autoclaved and normal soil in IF-I treatments at flowering indicates that in soil planted with rice, the O_2 released from the rice roots may also be supporting nitrification along with that produced in the upper oxygenated soil layer. Transpiration of the rice plants causes mass flow of water, resulting in mass flow of NH_4^+ as well toward the roots, supporting nitrification even under the anaerobic soil layer. As indicated earlier, a combination of NH_4^+ and NO_3^- leads to higher yields, greater by 40–70%, than does provide the same amount of N only as NH_4^+ [18]. Therefore, it appears that higher root activity for a longer duration, especially at the grain-filling stage, may help plants to get more of both forms of nitrogen even under flooded conditions, by supporting higher nitrification through the better supply of oxygen to the rhizosphere. This could be another reason for getting higher grain yield under the IF-I water regime than the IF-II and NF water regimes.

The study further revealed that the kinetics of available *N* and NH_4^+ was somewhat different in nonflooded soil (NF). The total available *N* was similar to that of IF-I, but it was present mainly in the form of NO_3^-. At the flowering stage, there was no effect of soil treatments on the available *N* content in the NF treatment. However, at 20 DAF, the EMS treatment had less *N* than AUS and NS treatments. Some case studies have demonstrated that the nitrogen requirement of microorganisms that decompose organic matter in aerated soils is higher than for decomposers in flooded soils, which results in higher net N immobilization in aerobic soils than in flooded soils [19]. This might be the reason for low soil N status in the EMS treatments compared to AUS and NS and so the lower root activity and early senescence. However, the biomass production was similar in all soil types under the NF water regimes (see **Table 1**), and the highest grain weight was recorded in the EMS and AUS treatments who received IF-I treatment.

Further, it can be seen that although there was no limitation of soil nitrogen in the NF water regimes, still plant biomass was not as significant as seen with IF-II. The possible reason could be a slower growth rate during the vegetative stage and lower cytokinin content in the roots. It is known that cytokinin content is regulated by soil nitrogen content and that the production of cytokinin as well as biomass is stimulated by having mixed source of nitrogen rather than only single source.

6. Discussions

6.1 Morphological and physiological plasticity of root architecture

Root length density—an important parameter of root morphology reflecting root architecture—is known to influence not only root-microbial interaction but also the physiological activity of roots, which plays an important role in increasing plants' photosynthetic capacity [20, 21]. Researchers have demonstrated that rice plants with higher root-oxidizing activity rate during their later growth stages have higher grain yield [22]. However, these findings were derived from rice plants with hybrid and "super" rice varieties which are known to have greater root activity than any traditional varieties [23].

Our preliminary studies [17] showed the significant effects of management practices such as intermittent irrigation or nonflooded water regimes on root development. The root architecture—defined here as root length density—significantly changed with mild water deficit. The response was not just at morphological level, but root activity also changed due to the effect of water and soil-available nitrogen and consequently also affected yield contributing parameters and finally grain yield. The root activity was higher in those plants who had higher chlorophyll content in their lower leaves at the later growth stage. Indeed this was related to high soil nitrogen content at later growth stage.

Many reports suggest that exploitation of soil resources through root activity may consume more than half of the available photosynthate in mature plants [24]. Given competing demands for internal plant resources for photosynthesis, support, defense, and reproduction, it is reasonable to expect that plastic response has favored plants that directed root activity to exploit efficiently, i.e., with a favorable balance of resource investment versus resource acquisition.

Knowing the effect of soil microbial density on soil nitrogen status, on important root traits under alternative soil water regimes, and the resulting effects on plant growth and performance helped to clarify the adaptive physiological response of plants under such different conditions.

It appeared that the combination of higher root-oxidizing activity rate, higher availability of NH_4^+/NO_3^- nitrogen, and higher chlorophyll content of the lower leaves at the later growth stage was one of the reasons for having higher yield under the two water conditions, IF-II and NF, compared to IF-II.

But we also noted that plants grown in autoclaved soil, either with IF-I or NS, had higher root activity rates than the other soil treatments. This increment did result in higher grain yield than with the other soil conditions; however, even with EM application, the root activity rate at a later growth stage was reduced significantly, but grain yield was similar to that of AUS soil treatments.

It seems that this physiological response of roots, i.e., their root activity rate, depends on the relative costs and benefits to the plant. If the supply of photosynthate to the roots, which comes mostly from the lower leaves of the plant, is restricted, or if the soil is limited in its nutrient availability and roots are unable to supply sufficient nutrients to the aboveground parts, the plasticity of response of plants' roots—either morphological proliferation or higher physiological activity—will be a burden for the plant.

Ultimately, the cost to the plant will depend on what is actually limiting its growth, whether nutrients or photosynthate supply. Therefore, the physiological basis of the plasticity of root and shoot growth needs to be understood inclusively within the context of environmental variables they are encountering with.

These works were the preliminary investigation and warrant further investigation at field level under different soil and weather conditions. However, the initial findings clearly showed that root architecture and root activity is greatly influenced by soil environment, particularly by water and soil microbial conditions. This flexibility arises due to the modular structure of roots which enables root deployment in zones rich in water and nutrients. The genetic control on this root deployment is still largely unknown, although the gene ANR1 is involved in the first stages of the nitrate (NO_3^-) signaling system when NO_3c- levels are locally enhanced [25]. This needs to be further studied under the subject of epigenetics.

6.2 Root plasticity under intermittent irrigation and opportunities for mitigation of methane production in the rice field

While there is a need to continue research to identify and/or induce more productive genotypes in general, concern for dealing with climate change should prompt more research particularly on how best to modify crop management to take advantage of plants' inherent plasticity of morphological and physiological response to environmental influences that would otherwise be limiting factors and constraints.

It is known that up to 90% of the CH_4 emitted in rice paddies is released through rice transport [26], while between 19 and 90% of the CH_4 produced is oxidized, with up to 75% of the CH_4 oxidation taking place in the rhizosphere [27]. Accordingly, strategies to lower net CH_4 emission from rice fields include reduction of CH_4 production, increasing CH_4 oxidation, and lowering CH_4 transport through the plant. Among the CH_4 emission mitigation strategies that do not compromise rice productivity, the introduction of drainage periods during the crop cycle appears to be the most efficient [28]. Thus, it has been estimated that intermittent drainage periods by applying intermittent irrigation in poorly drained rice fields could reduce 10% the agricultural CH_4 emissions [29].

It is expected that the higher root activity rate for a longer duration, as appeared in our studies, should further enhance CH_4 oxidation in the rhizosphere because of the prolonged oxygenated rhizosphere. This benefit will be relatively higher under intermittent irrigation water regimes, but even under flooded condition, a relative mitigation benefit can be achieved through minimizing intra-hill competition since minimizing intra-hill competition can also enhance root activity [30].

In the present study, aerobic soil was maintained for some period in IF-I, IF-II, and for the whole crop growth period in the NF water regime. It could be assumed that under continuously flooded water regimes, the soil would be anaerobic except 2–5 cm depth from the surface of the soil. But even under this condition, the root length density was better at 15–20 cm soil depth at the flowering stage (**Figure 1**). It shows that oxygen concentration required for the development of laterals was present in this zone even under continuously flooded water regimes.

The earlier findings suggest that for an aerobic rhizosphere, spacing is critical along with the number of primary roots per plant [31]. For example, if the number of primary roots is 500, and the hill spacing is 25 × 25 cm, then the numbers of root/cm^2 = 0.8 root cm^{-2}. Thus for FO_2 A_R (where FO_2 = flux of oxygen across root surface, and A_R = surface area of roots capable of absorption) = 0.2 nmol s^{-1} (which is standard rate under flooded condition), the rate of release of oxygen will be 160 pmol cm^{-2} (soil surface) s^{-1}. This amount of oxygen is sufficient for the growth of laterals as well as nitrogen uptake by the plants in the form of ammonium and nitrate under flooded condition.

Typically, the maximum rate of nitrogen uptake by rice crop are ≤5 kg h^{-1} day^{-1} [32] or 40 pmol cm^{-2} (soil surface) s^{-1}. Therefore, if half the oxygen released from the roots was used to nitrify ammonium in the rhizosphere (NH_4^+ + $2O_2 \rightarrow NO_3^- + 2H^+ + H_2O$), and half the nitrate produced was recovered by the roots, an oxygen release of 160 pmol cm^{-2} (soil surface) s^{-1} would be sufficient to nitrify half the nitrogen by the roots and also methane oxidation. This would facilitate uptake of nitrogen in the form of nitrate and ammonium as well for higher biomass production along with methane emission reduction from paddy fields. Therefore, aerobic rhizosphere can be maintained even under shallow flooded condition by minimizing intra-hill competition, by transplanting fewer seedlings/hill with wider spacing.

In addition, intermittent irrigation or keeping soil "preferably moist" or in nonflooded condition will reduce aerenchyma formation rate. Since the aerenchyma acts as a channel for oxygen transport from the atmosphere to the roots and CH_4 transport from the site of production to the atmosphere, therefore, reduced aerenchyma formation will lead to lowering CH_4 transport through the plant.

These benefits become more relevant in the prospective scenario where rice production needs to be increased with both reduced water applications and reduced "climate-forcing" practices.

These initial findings are opening up many possibilities for better understanding of plants' growth response and root plasticity under varied soil environments which could be exploited and manipulated to enhance crop production through enhanced root/rhizosphere activity.

Since, agronomic crop management practices (avoiding continuous soil saturation, minimizing intra-hill competition, applying effective microorganism, organic manure, aerating the soil, etc.) are seen to increase root growth and yield from practically any variety. Our research suggests that positive responses can be induced through appropriate water management practices and with an increased microbial density that can increase the total root and shoot growth and plant biomass. It also suggests that the roots and shoots are not necessarily in a zero-sum relationship, as posited by harvest index thinking; with appropriate agronomy, there can be positive feedbacks between each, as evident from this study. Therefore, such management practices should be explored in detail to gain a better understanding of root and rhizosphere activity.

7. Conclusions

Climate change is altering the growing environments for plants, particularly aboveground, but there are also belowground effects as changes in precipitation and in ambient temperature have a strong influence on soil conditions. Plant species are genetically programmed to adjust to the novel conditions through phenotypic plasticity. While there is a need to continue research to identify and/or induce more productive genotypes in general, concern for dealing with climate change should prompt more research particularly on how best to modify crop management to take advantage of plants' inherent plasticity of morphological and physiological response to environmental influences that would otherwise be limiting factors and constraints.

Our results and discussion document that rice root morphology and physiology and consequently rice shoot growth are significantly affected by variations in soil water conditions. Root architecture and roots' oxidizing activity rate are important factors--influencing higher yield--are quite plastic in nature and vary considerably with varying water regimes and with varying soil microbial population. Modifying water management to take advantage of plants' inherent plasticity of morphological and physiological response can be one of the adaptive strategies for achieving higher yield under reduced water condition along with mitigation of methane production from rice fields. Such an investigation would be useful to develop alternative crop management practices that will reduce "climate forcing" and will provide better ecosystem services.

Conflict of interest

The authors declare no conflict of interest.

Author details

Abha Mishra
ACISAI Center, Asian Institute of Technology, Pathum Thani, Thailand

*Address all correspondence to: abhamishra@ait.asia

References

[1] Gewin V. Food: An underground revolution. Nature. 2010;**466**:552-553

[2] Stearns SC. The evolutionary significance of phenotypic plasticity. Bioscience. 1989;**39**:436-445

[3] Garg KP, Welch LF. Growth and phosphorus uptake by corn as influenced by phosphorus placement. Agronomy Journal. 1967;**59**:152-154

[4] Yao J, Barber SA. Effect of one phosphorus rate placed in different soil volumes on P uptake and growth of wheat. Communication in Soil Science and Plant Science. 1986;**17**:819-827

[5] Anghinoni I, Barber SA. Predicting the most efficient phosphorus placement for corn. Soil Science Society of America Journal. 1980;**44**:1041-1044

[6] Samejima H, Kondo M, Ito O, Nozoe T, Shinano T, Osaki M. Characterization of root system with respect to morphological traits and nitrogen absorbing ability in new plant type of tropical lines. Journal of Plant Nutrition. 2005;**50**:545-554

[7] Hodge A, Berta G, Doussan C, Merchan F, Crespi M. Plant root growth, architecture and function. Plant and Soil. 2009;**321**:153-187

[8] Bouman BAM, Lampayan RM, Tuong TP. Water Management in Irrigated Rice: Coping with Water Scarcity. Los Baños, Philippines: International Rice Research Institute; 2007. 54 pp

[9] McCauley GN. Sprinkler vs. flood irrigation in traditional rice production regions of Southeast Texas. Agronomy Journal. 1990;**82**:677-683

[10] Mishra HS, Rathore TR, Pant RC. Root growth, water potential and yield of irrigated rice. Irrigation Science. 1997;**17**:69-75

[11] Sah RN, Mikkelsen DS. Availability and utilization of fertilizer nitrogen by rice under alternate flooding. Plant and Soil. 1983;**75**:221-226

[12] Ceesay M, Reid SW, Fernandes ECM, Uphoff NT. The effect of repeated soil wetting and drying on low land rice yield with system of rice intensification (SRI) methods. International Journal of Agricultural Sustainability. 2006;**4**:5-14

[13] Sato S, Uphoff NT. Raising factor productivity in irrigated rice production: Opportunities with the system of rice intensification. CAB review: Perspectives in Agriculture, Veterinary Science, Nutrition and Natural Resources. Wallingford, UK: Commonwealth Agricultural Bureau International; 2007;54:2

[14] Yang J, Zhang J, Wang Z, Liu L, Zhu O. Post-anthesis water deficit enhance grain filling in two-line hybrid rice. Crop Science. 2003;**43**:2099-2108

[15] Yang J, Zhang J. Crop management techniques to enhance harvest index in rice. Journal of Experimental Botany. 2010;**61**:3177-3189

[16] Tennant D. A test of modified line intersected method of estimating root length. Journal of Ecology. 1975;**63**:995-1001

[17] Mishra A, Uphoff NT. Morphological and physiological response of rice root and shoots to varying water regimes and soil microbial densities. Archives of Agronomy and Soil Science. 2013;**59**:705-731

[18] Kronzucker HJ, Schjoerring JK, Erner Y, Kirk GJD, Siddiqi MY, Glass ADM. Dynamic interactions

between root NH_4+ influx and long-distance N translocation in rice: Insights into feedback processes. Plant and Cell Physiology. 1999;**39**:1287-1293

[19] Broadbent FE. Mineralization of organic nitrogen in paddy soils. In: Brady NC, editor. Nitrogen and Rice. Laguna, Philippines: Los Baños; 1979. pp. 105-118

[20] Mitsui S, Tensho K. Dynamic studies on the nutrient uptake by crop plants. 3: The reducing power of roots of growing plants as revealed by nitrite formation in the nutrient solution. Journal of the Science of Soil and Manure (Japan). 1953;**22**:301-307

[21] Ookawa T, Naruoka Y, Yamazaki T, Suga J, Hirasawa T. A comparison of accumulation and partitioning of nitrogen in plants between two rice cultivars (Akenohoshi and Nipponbare) at the ripening stage. Plant Production Science. 2003;**6**:172-178

[22] Osaki M, Shinano T, Matsumoto M, Zheng T, Tadano T. A root-shoot interaction hypothesis for high productivity of field crops. Soil Science and Plant Nutrition. 1997;**43**:1079-1084

[23] Ookawa T, Naruoka Y, Sayama A, Hirasawa T. Cytokinin affects the leaf levels of ribulose-1,5-bisphosphate carboxylase/oxygenase and the accumulation and partitioning of nitrogen in paddy rice at the ripening stage. Crop Science. 2004;**44**:2107-2115

[24] Fogel R. Roots as primary producers in below-ground ecosystems. In: Fitter AH, Atkinson D, Read DJ, Usher MB, editors. Ecological Interactions in Soil: Plant Microbes and Animals, Special Publication 4 of the British Ecological Society. Oxford, UK: Blackwell Scientific Publications; 1985. pp. 23-36

[25] Zhang H, Forde BG. An Arabidopsis MADS box gene that controls nutrient-induced changes in root architecture. Science. 1998;**279**:407-409

[26] Conrad R. Microbial ecology of methanogens and methanotrophs. Advances in Agronomy. 2007;**96**:1-63

[27] Frenzel P. Plant-associated methane oxidation in rice fields and wetlands. Advanced Microbial Ecology. 2000;**16**:85-114

[28] Neue HU. Methane emission from rice fields. Bioscience. 1993;**43**:466-474

[29] Kern JS, Gong ZT, Zhang GL, Zhuo HZ, Luo GB. Spatial analysis of methane emission from paddy soil in China and the potential for emission reduction. Nutrient Cycling in Agroecosystems. 1997;**49**:181-195

[30] Mishra A, Salokhe VM. The effects of planting pattern and water regime on root morphology, physiology, and grain yield of rice. Journal of Agronomy and Crop Science. 2010;**196**:368-378

[31] Kirk GJD. Rice root properties for internal aeration and efficient nutrient acquisition in submerged soil. The New Phytologist. 2003;**159**:185-194

[32] Peng SB, Cassman KJ. Upper thresholds of nitrogen uptake rates and associated nitrogen fertilizer efficiencies in irrigated rice. Agronomy Journal. 1998;**90**:178-185

Chapter 3

Nitrogen Transport in Barley

Salwa Abdel-latif, Hanan Abou-Zeid and Kuni Sueyoshi

Abstract

The translocation of nitrate in intact plant of barley (*Hordeum vulgare* L. cv.) two genotypes, wild type Steptoe and a mutant Az12, was visualized by a positron-emitting tracer imaging system (PETIS) after supplying positron-emitting ^{13}N-labelled nitrate ($^{13}NO_3^-$) to the seedlings. ^{13}N movement was monitored to visualize the distribution of radioactivity in the two genotypes. N sufficient seedlings causes enhanced N uptake and translocation to shoots in time course from (0, 10, 20, 30, 40 min). The N-depleted seedlings were exposed to a nutrient solution containing nitrate and nitrite, and were labeled with ^{15}N for 38 h under (14L/10D) cycles. The two genotypes utilized $^{15}NO_3^-$ and accumulated it as reduced ^{15}N, predominately in the shoots. In the Az12, nitrate accumulation in shoots was 78% higher than that in the Steptoe. Accumulation of reduced ^{15}N in the Az12 roots was nearly similar to that of the Steptoe roots, but 8% lower in the Az12 shoots than in the Steptoe shoots at the end of the experiment.

Keywords: barley, mutant, nitrate reduction, light/dark, ^{15}N incorporation model, PETIS—a positron-emitting tracer imaging system

1. Introduction

The two barley genotypes were specifically chosen since they differ only in the distribution of nitrate reductase (NR). Imaging technologies using high-energy emitting radio isotopes and radionuclide tracers allow researchers to visualize the dynamics (absorption, translocation and distribution) of mineral movement in plant, understand the dynamics of water, nutrient, pollutants in plants and to analyze the plant physiology of a test plant. The use of γ-rays emitted from positrons in $^{11}CO_2$ [1–3] ^{13}NH or ^{13}NO [4, 5] and $H_2{}^{18}F^-$ [6] has been adopted in plant nutrition research. Among these technologies, the positron-emitting tracer imaging system (PETIS) [7] which was designed for studying plant physiology and agriculture, has often been used to examine the distribution and translocation of nutrients using positron-emitting tracers. Several positron emitting radioisotopes such as ^{11}C and ^{13}N can be used in plant biology research. In general, radioisotope tracers are useful tools for analyzing the spatial distribution or temporal change in the amount of a substance in the plant body.

The PETIS was recently used to visualize the accumulation of photo assimilates in grains of a wheat ear [8] with 2.3 mm resolution. In recent years, the PETIS has been employed to study various physiological functions in intact, living plants [9, 10] one of the most advanced radiotracer-based imaging methods available today. This system enables not only monitoring of the real-time movement of the tracer in living plants as a video camera might, but also quantitative analyses of the movement of the substance of interest by freely selecting a region of interest on the image data obtained. Short-lived radioisotope techniques provide data that are crucial for developing models that quantitatively link the underlying biochemical reactions to physiological responses.

2. Visualization of ^{13}N accumulation in barley using (PETIS)

2.1 ^{13}N analysis in Steptoe

Short incubation times are used in order to determine the fate of nitrate transported in the shoot of two barley genotypes. Therefore, $^{13}NO_3^-$ was applied to barley seedling and short-term ^{13}N distribution was determined using PETIS. The imaging pictures of ^{13}N radioactivity were monitored after $^{13}NO_3^-$ was supplied to the medium containing 2.0 mM KNO_3 PETIS images show high ^{13}N accumulation in Az12 shoot than in Steptoe shoot (**Figure 1**).

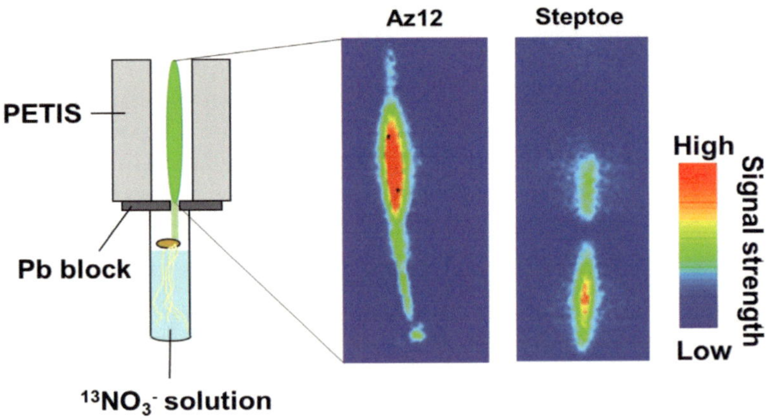

Figure 1.
Imaging of radioactivity in barley shoot supplied with $^{13}NO_3^-$ for 40 min using PETIS.

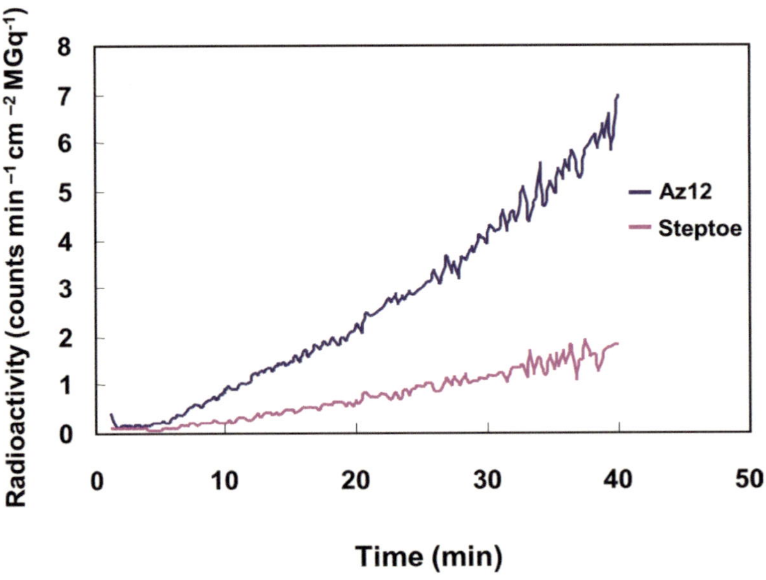

Figure 2.
Changes in the relative counts radioactivity for 40 min in barley shoot of wild type Steptoe and mutant Az12 measured by PETIS.

The $^{13}NO_3^-$ was supplied to barley seedlings for the two genotypes and the translocation of $^{13}NO_3^-$ were monitored using PETIS. The seedlings were incubated with 2.3 mM KNO_3^-, one seedling was selected and transferred to a feeding container that contained 8 mL of pretreatment solution. Subsequently, 100 μL of 2.3 mM KNO_3^- and 2 mL of $^{13}NO_3^-$ solution (113 MBq) were immediately added (final concentration of KNO_3^- is 2.0 mM).

A plot of relative radioactivity was shown in **Figure 2**. The radioactivity was shown at 40 min after the addition of $^{13}NO_3^-$ containing 2.0 mM KNO_3^- in the Steptoe and Az12 plants. The radioactivity linearly increased over time after a lag of several minutes for both genotypes. The count of radioactivity of Az12 shoots at 40 min after $^{13}NO_3^-$ supply was about four times higher than that of Steptoe. These results suggest that the excess nitrate accumulation in Az12 plants shoots is probably due to the lower capacity of the mutant to reduce nitrate [11, 12].

3. Nitrate reduction and allocation of reduced nitrogen in roots and shoots of the wild type Steptoe and the mutant Az12 using ^{15}N-tracing method to determine accumulation, uptake, translocation and reduction of nitrate, together with transport of reduced ^{15}N in intact N-starved seedlings under light/dark cycles

Barley besides its importance as a crop is an established model plant for both genetic and physiological studies [13]. In the two barley genotypes (*Hordeum vulgare* L.) wild type Steptoe and the mutant Az12 using ^{15}N-tracing method to determine accumulation, uptake, translocation and reduction of nitrate, together with transport of reduced ^{15}N in intact N-starved seedlings under light/dark cycles. Also for both genotypes root contribution increased during L/D cycles and decreased during the subsequent light cycle. Shoot nitrate accumulation in Az12 was higher than in Steptoe. Nitrate-deficient barley seedlings showed negligible accumulation of short-lived tracer $^{13}NO_3$- in shoots than did N-sufficient barley and in Az12 more than in Steptoe genotypes revealing that the N sufficient seedlings caused enhancement of nitrate uptake and translocation to shoots [14]. Barley is a highly adaptable cereal grain and ranks 5th among all crops for dry matter production in the world since it is an important food source of protein in many parts of the world. Nitrogen is considered as one of the three macronutrients required for high crop yields and has a critical role in plant growth and development [15, 16]. Three quarters of our atmosphere consists of nitrogen gas (N_2) and elemental nitrogen must be transformed to usable forms before it is available for plant uptake. Addition of NO_3^- to the incubation medium of dark-grown NO_3^--starved Steptoe seedlings resulted in a greater accumulation of NO_3^- in leaves and roots more than those of illuminated-grown seedlings [17]. Plants require more nitrogen than any other nutrient [18]. Nitrogen is an important component of various compounds such as amino acids, amides and proteins, quaternary ammonium compounds and polyamines [19–21]. Nitrate (NO_3^-) is the major N-source for cultivated plants and is an important signaling ion that influences plant growth and differentiation [22]. NO_3^- uptake, transport, and responses have been a major focus of research. In addition to its role as a nutrient, NO_3^- can act as a signaling molecule that modulates gene expression and a wide range of processes including plant growth, root system architecture [23, 24]. Roots are crucial for perception and uptake of nitrate in plants [25–27].

The first enzymatic step of nitrate after its active transport into the cell, is the reduction to nitrite which achieved by nitrate reductase (NR). In higher plants NR exists in two forms: NADH:NR (E.C 1.6.6.1) and NAD(P)H:NR (E.C.1.6.6.2). The bispecific NAD(P)H:NR is found in the non-green tissues of monocotyledons and in both green and non-green tissues of legume. NR is a key enzyme in a plant's nitrogen assimilation pathway, this step of the reduction often considered to be rate limiting step and nitrite transfers to plastids where it quickly reduced to NH_4^+ by NiR enzyme [28]. In a second step, nitrite reductase which is localized in the chloroplast, catalyses the reduction of nitrite to ammonia before incorporation into amino compounds so, for protein synthesis [29–31]. This reduction can take place in either roots or leaves, depending on plant species, age and nitrate supply rate. It has been reported that the factors which regulate nitrate reductase enzyme are inorganic salts and ions, antibiotics and metabolic inhibitor, fungicides and herbicides, seedling age and diurnal rhythms, temperature, water stress and gaseous environment, atmospheric pollutants and external pH [32]. For higher plants, nitrate is the major source of inorganic nitrogen, which is translocated to the leaf, and assimilated and metabolized into various organic compounds utilizing reductant provided by photosynthesis. The reduced nitrogen compounds are incorporated into various biomacromolecules, such as proteins and nucleic acids. Nitrate uptake and reduction are considered the initial processes by which NO_3^- is metabolized by higher plants, are modulated by light and dark [33] and also of interest in understanding plant nitrogen nutrition and the plant nitrate assimilation pathway. Nitrogen assimilation is a fundamental biological process that has a marked effect on plant productivity, biomass and crop yield. It is well established that plants supplied with excess nitrate of current demand have the ability to accumulate nitrate. The manner in which nitrate stored or reduced and assimilated in both roots and shoots depends on plant species [34]. Depending on the nitrogen demand, nitrate is directed into several routes after uptake into the root cells: it can be translocated to the shoot, stored in the vacuole, or added to the cytosolic pool. Some aspects of plant nitrogen metabolism have been studied in detail in barley [5, 35, 36]. Several studies with mutants or trasformants with altered NR expression clearly showed that there is no direct correlation between plant growth and the nitrate reduction capacity of the plant also showed a reverse relation between nitrate content and NR activity, i.e., plants with decreased NR activity contain more nitrate and are phenotypically not different. It is only when NR activity is decreased below 10% of the wild type level is plant growth and protein affected [37–39]. Both roots and shoots of wild-type Steptoe contain the two characterized isozymes of NR (NADH-specific and NAD(P)H-bispecific) typically found in barely, where the NAD(P)H-specific NR is not expressed in leaves, but is induced by nitrate in roots. In Az12, NAD(P)H-specific NR is present in roots and shoots, since Az12 is a mutant that is affected by partial or complete loss of its capacity to either reduce nitrate or produce nitrite under NR assay conditions, yet has low levels of NR activity by some unknown nitrate assimilatory pathway [11].

Nitrate assimilation is the primary pathway by which plants obtained reduced nitrogen. In many species of higher plants most organic N is derived from the assimilation of nitrate in the shoots [12, 40]. Short-time labelings are generally used by several authors [41–44] to prevent the minor allocation of reduced ^{15}N onto and out of the roots. Most studies showed that when plants are exposed to nitrate a continuous increase of nitrate uptake was achieved by the roots. However long time exposure to nitrate may inhibit nitrate uptake [45]. Use of Az12 mutant provide a simpler system to study the characteristics of nitrate reduction since the two genotypes are phenotypically the same but differ only in the distribution of NR [46]. For that the whole plant contribution to nitrate reduction occurred upon the early stages of N utilization when the induction is not fully achieved.

Since many studies on the physiological characterizations of over expression and under expression NR genotypes include no measurements of the in vivo nitrate reduction rate. The fact that using whole plant in experiments to investigate the contribution between shoots and roots to reduced ^{15}N is complicated although it has an advantage for measuring the assimilation and translocation of reduced N in intact seedlings and consequently to the measurement of the whole plant nitrate reduction. Since some limitations and disadvantages may be involved as a result of using

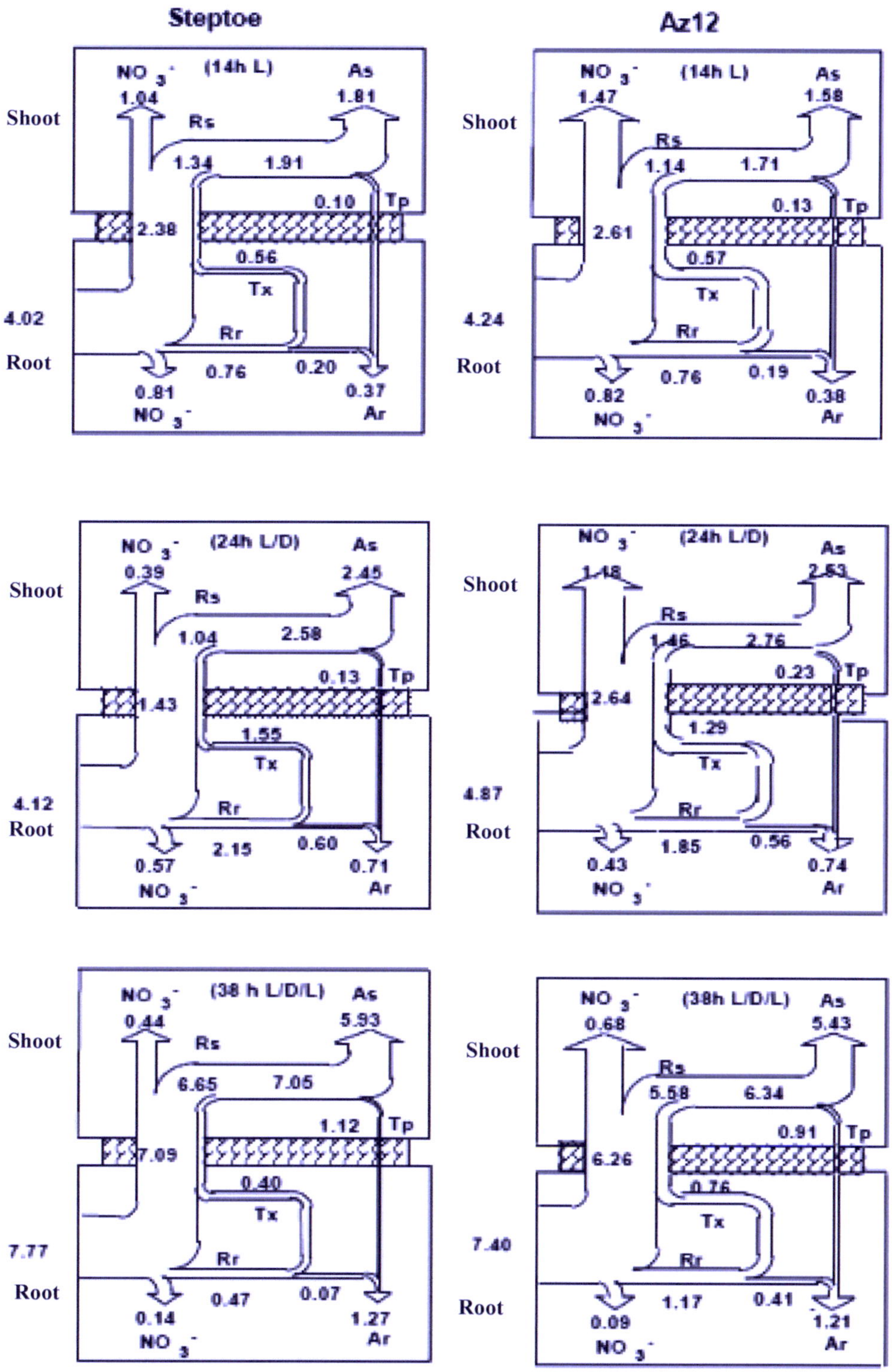

Figure 3.
Balance sheets for uptake, accumulation, reduction, and translocation of nitrate, and accumulation and translocation of reduced ^{15}N in barley seedlings.

excised tissues [47]. The present study was conducted to investigate the differences between the wild-type (Steptoe) and the mutant (Az12) in nitrate reduction, uptake and the transport of reduced ^{15}N between roots and shoots using ^{15}N labeling in a split root experiment. Also, to study the upward and downward translocation of reduced ^{15}N in intact barley seedlings which were assimilating nitrate from a mixed N-medium, by using the ^{15}N incorporation model [48, 49]. The mutant Az12 plants deficient in NADH-specific NR was used with wild-type Steptoe as a control.

Nitrate reduction during the first 14 h light period accounted for 54 and 46% of total nitrate absorbed by the plants, respectively, in Steptoe and Az12. In both genotypes, nitrate accumulation in root was occurred mainly during the first 14 h light period and accounted for about 19% of the total amount of absorbed nitrate. In the other hand, reduced ^{15}N accumulation in roots was low and quit constant during whole period as amounted to about 8% of total absorbed nitrate. Shoot nitrate reduction during the first 14 h light period accounted for 33 and 27% of total nitrate uptake, respectively, in Steptoe and Az12, and for 25 and 30% during the dark period, and for 86 and 75% during the second light period. In both genotypes about 60% of absorbed nitrate was transported to the shoot via xylem during the first 14 h light period and the proportion decreased during dark period, then increased again at the subsequent light period and this may be due to light/dark transition. From the same reason also fluctuation was observed in both shoot reduced ^{15}N accumulation and shoot nitrate reduction in Steptoe where at first 14 h light period, shoot nitrate reduction accounted for 33, 25 and 86% of total nitrate uptake, respectively, in light/dark/light periods. On the other hand, shoot nitrate reduction in Az12 accounted for 27, 30 and 75% of total nitrate, respectively, in light/dark/light periods. Nitrate translocation to shoots at 14 h light for both genotypes was 59% in Steptoe and 62% in Az12 of total nitrate uptake then decreased to about 34% for both genotypes at dark period and increased to 91% in Steptoe and 75% in Az12 at subsequent second light period (**Figure 3**).

Az12 was used as a tool in order to assess the importance of root and shoot nitrate reduction and allocation of reduced N for the N-nutrition of barley plants. Although some differences between Az12 and Steptoe in this study, it could be concluded that the overall fate of the absorbed nitrate was basically similar between the two genotypes under light/dark cycle.

4. Nitrate accumulation

The accumulation of nitrate in barley (*Hordeum vulgare* L. cv. Giza 123) changed in roots and leaves at light/dark 14:10 h cycle (**Table 1**). The root nitrate content was 0.65–1.68 μmol plant^{-1} respectively during the 14 h light and 10 h dark periods after the transfer to nutrient solution containing nitrate. In the dark nitrate content of both roots and leaves was elevated. However, the nitrate content decreased in roots

Plant part	14 h (light)	24 h (light-dark)	38 h (light-dark-light)
Roots	0.65 ± 0.06	1.68 ± 0.13	1.12 ± 0.15
Shoots	0.79 ± 0.11	1.23 ± 0.11	0.53 ± 0.17
Total plant	1.44 ± 0.14	2.91 ± 0.23	1.65 ± 0.27

N-depleted seedlings (9-d-old) were treated with a nutrient solution containing 2.5 mM KNO_3^- under a light-dark cycle of 14:10 h at 25°C. Leaves and roots were harvested after 14, 24, and 38 h in the nutrient solution. Results are the means of three replicates (3 × 5 plants) ± SE.

Table 1.
Nitrate accumulation in roots and leaves of barley seedlings after treatment with a N-medium.

and leaves during the 38 h light period and significantly low nitrate contents were measured in roots and leaves. On a whole plant basis, roots accounted for 45 (14 h), 58 (24 h) and 68% (38 h) of the nitrate accumulation of the whole plant during light/dark/light period [17].

Acknowledgements

I would like to thank Professor Dr. Takuji Ohyama at Department of Applied Biological Chemistry, Faculty of Agriculture, Niigata University, Japan, for their technical assistance and help during the experimental work.

Conflict of interest

No potential conflict of interest was reported by the authors.

Abbreviations

Anl	accumulation of reduced ^{15}N from $^{15}NO_3^-$ in non-labeled roots of split roots
Ar	accumulation in roots of reduced ^{15}N from $^{15}NO_3^-$
As	accumulation in shoots of reduced ^{15}N from $^{15}NO_3^-$
Rr	$^{15}NO_3^-$ reduction in roots
Rs	$^{15}NO_3^-$ reduction in shoots
Tp	translocation to root of shoot reduced ^{15}N from $^{15}NO_3^-$ in phloem
Tx	translocation to shoot of root-reduced ^{15}N from $^{15}NO_3^-$ in xylem
PETIS	a positron-emitting tracer imaging system

Author details

Salwa Abdel-latif[1*], Hanan Abou-Zeid[2] and Kuni Sueyoshi[3]

1 Faculty of Education, Alexandria University, Alexandria, Egypt

2 Faculty of Science, Alexandria University, Alexandria, Egypt

3 Faculty of Agriculture, Niigata University, Niigata, Japan

*Address all correspondence to: salwa_latif@yahoo.com

References

[1] Fritz R, Wieneke J, Fuehr F. Transport kinetics of carbon-11 labelled photosynthates in soybeans. I. Effect of time of day and physiological age. Angew and te Botanik. 1986;**60**:317-332

[2] Grusak MA, Minchin PEH. Seed coat unloading in *Pisum sativum*: Osmotic effects in attached versus excised empty ovules. Journal of Experimental Botany. 1988;**39**:543-560

[3] Roeb G, Britz SJ. Short-term fluctuations of the transport of assimilates to the ear of wheat measured with steady-state carbon-11carbon dioxide labeling of the flag leaf. Journal of Experimental Botany. 1991;**42**:469-476

[4] Presland MR, McNaughton GS. Whole plant studies using radioactive ^{13}N-nitrogen. Journal of Experimental Botany. 1986;**37**:1619-1632

[5] Kronzucker HJ, Glass ADM, Siddiqi MY. Inhibition of nitrate uptake by ammonium in barley. Analysis of component fluxes. Plant Physiology. 1999;**120**:283-291

[6] McKay RM, Palmer GR, Ma XP, Layzell DB, McKee TA. The use of positron emission tomography for studies of long-distance transport in plants: Uptake and transport. Plant, Cell and Environment. 1988;**11**:851-861

[7] Mori S, Kiyomiya S, Nakanishi H, Ishioka NS, Watanabe S, Osa A, et al. Utilization of ^{15}O-water flow in tomato and rice in the light and dark using a positron-emitting tracer imaging system (PETIS). Soil Science & Plant Nutrition. 2000;**46**:975-979

[8] Matsuhashi S, Fujimakia S, Uchida H, Ishioka SN, Kume TA. A new visualization technique for the study of the accumulation of photo assimilates in wheat grains using [^{11}C] CO_2. Applied Radiation and Isotopes. 2006;**64**:435-440

[9] Fujimaki S. The positron emitting tracer imaging system (PETIS), a most-advanced imaging tool for plant physiology. ITE Letters on Batteries New Technologies & Medicine. 2007;**8**:404-413

[10] Kiser MR, Reid CD, Crowell AS, Phillips RP, Howell CR. Exploring the transport of plant metabolites using positron emitting radiotracers. Human Frontier Science Program. 2008;**2**:189-204

[11] Gojon A, Dapoigny L, Lejay L, Tillard P, Rufty TW. Effects of genetic modifications of nitrate reductase expression on $^{15}NO_3^-$ uptake and reduction in *Nicotiana* plants. Plant, Cell and Environment. 1998;**21**:43-53

[12] Warner RL, Huffaker RC. Nitrate transport is independent of NADH and NAD (P) H nitrate reductases in barley seedlings. Plant Physiology. 1989;**91**:947-953

[13] Forster BP, Ellis RP, Thomas WTB, Newton AC, Tuberosa R, This D, et al. The development and application of molecular markers for abiotic stress tolerance in barley. Journal of Experimental Botany. 2000;**51**:19-27

[14] Abdel-Latif S, Abou-Zeid HM. Visualization of ^{13}N accumulation in two cultivars of barley plants using a positron-emitting tracer imaging system (PETIS). Indian Stream Research Journal. 2013;**3**:1-4

[15] Krapp A, David LC, Chardin C, Girin T, Marmagne A, Leprince AS, et al. Nitrate transport and signalling in *Arabidopsis*. Journal of Experimental Botany. 2014;**65**:789-798

[16] Ruffel S, Gojon A, Lejay L. Signal interactions in the regulation of root nitrate uptake. Journal of Experimental Botany. 2014;**65**:5509-5517

[17] Abdel-Latif S, Abou-Zeid HM. Activation and deactivation of nitrate reductase isozymes in the leaves of barley plants under light and dark conditions. Life Science Journal. 2014;**11**(9):999-1004

[18] Premachandra D, Hudek L, Brau L. Bacterial modes of action for enhancing of plant growth. Journal of Biotechnology and Biomaterials. 2016;**6**:236-244

[19] Rais I, Masood A, Inam A, Khan N. Sulfur and nitrogen coordinately improve photosynthetic efficiency, growth and proline accumulation in two cultivars of mustard under salt stress. Journal of Plant Biochemistry and Physiology. 2013;**1**:101-107

[20] Zaki SS. Effect of compost and nitrogen fertilization on yield and nutrients uptake of rice crop under saline soil. Modern Chemistry and Applications. 2016;**4**:183-186

[21] Arghavani M, Zaeimzadeh A, Savadkoohi S, Samiei L. Salinity tolerance of Kentucky bluegrass as affected by nitrogen fertilization. Journal of Agricultural Science and Technology. 2017;**19**:173-183

[22] Kaiser WM, Huber SC. Post-translational regulation of nitrate reductase: Mechanism, physiological relevance and environmental triggers. Journal of Experimental Botany. 2001;**52**:1980-1989

[23] Krouk G, Crawford NM, Coruzzi GM, Tsay YF. Nitrate signaling: Adaptation to fluctuating environments. Current Opinion in Plant Biology. 2010;**13**:265-272

[24] Alvarez JM, Vidal EA, Gutierrez RA. Integration of local and systemic signaling pathways for plant N responses. Current Opinion in Plant Biology. 2012;**15**:185-191

[25] Mounier E, Pervent M, Ljung K, Gojon A, Nacry P. Auxinmediated nitrate signaling by NRT1. 1 participates in the adaptive response of *Arabidopsis* root architecture to the spatial heterogeneity of nitrate availability. Plant, Cell and Environment. 2014;**37**:162-174

[26] Léran S, Edel KH, Pervent M, Hashimoto K, Corratgé-Faillie C, Offenborn JN, et al. Nitrate sensing and uptake in *Arabidopsis* are enhanced by ABI2, a phosphatase inactivated by the stress hormone abscisic acid. Science Signaling. 2015;**8**:ra43. DOI: 10.1126/scisignal.aaa4829

[27] Kiba T, Krapp A. Plant nitrogen acquisition under low availability: Regulation of uptake and root architecture. Plant and Cell Physiology. 2016;**57**:707-714

[28] Rosales EPMF, Iannone M, Groppa DM, Benavides P. Nitric oxide inhibits nitrate reductase activity in wheat leaves. Plant Physiology and Biochemistry. 2011;**49**:124-130

[29] Crawford NM. Nitrate: Nutrient and signal for plant growth. The Plant Cell. 1995;7:859-868

[30] Mane AV, Deshpande TV, Wagh VB, Karadge BA, Samant JSA. Critical review on physiological changes associated with reference to salinity. International Journal of Environmental Sciences. 2011;**6**:1192-1216

[31] Zhang YL, Zhang L, Hu XH. Exogenous spermidine-induced changes at physiological and biochemical parameters levels in tomato seedling grown in saline-alkaline condition. Botanical Studies. 2014;**55**:55-58

[32] Bose B, Srivastava HS. Absorption and accumulation of nitrate in plants: Influence of environmental factors. Indian Journal of Experimental Biology. 2001;**39**:101-110

[33] Aslam M, Huffaker RC, Rains DW, Rao KP. Influence of light and ambient carbon dioxide concentration on nitrate assimilation in intact barley seedlings. Plant Physiology. 1979;**63**:1205-1209

[34] Marschner H. Mineral Nutrition of Higher Plants. London: Academic Press; 1995

[35] Botrel A, Kaiser WM. Nitrate reductase activation state in barley roots in relation to the energy and carbohydrate status. Planta. 1997;**201**:496-501

[36] Vidmar JJ, Zhuo D, Siddiqi MY, Schjoerring JK, Touraine B, Glass ADM. Regulation of high-affinity nitrate transporter genes and high-affinity nitrate influx by nitrogen pools in roots of barley. Plant Physiology. 2000;**123**:307-318

[37] Wilkinson JQ, Crawford NM. Identification and characterization of a chlorate-resistant mutant of *Arabidopsis thaliana* with mutations in both nitrate reductase structural genes NIA1 and NIA2. Molecular and General Genetics. 1993;**239**:289-297

[38] Foyer CH, Lescure JC, Lefebvre C, Morot GJF, Vincentz M, Vaucheret H. Adaptations of photosynthetic electron transport, carbon assimilation and carbon partitioning in transgenic *Nicotiana plumbaginifolia* plants to changes in nitrate reductase activity. Plant Physiology. 1994;**104**:171-178

[39] Scheible WR, González-Fontes A, Morcuende R, Lauerer M, Geiger M, Glaab J, et al. Tobacco mutants with a decreased number of functional nia genes compensate by modifying the diurnal regulation of transcription, post-translational modification and turnover of nitrate reductase. Planta. 1997;**203**:304-319

[40] Beevers L, Hageman RH. Nitrate and nitrite reduction. In: Miflin BJ, editor. The Biochemistry of Plants, Amino Acids and Derivatives. Vol. 5. New York: Academic Press; 1980. pp. 115-168

[41] Ashley DA, Jackson WA, Volk RJ. Nitrate uptake and assimilation by wheat seedlings during initial exposure to nitrate. Plant Physiology. 1975;**55**:1102-1106

[42] Talouize AG, Guiraud G, Moyse A, Marol C, Champigny ML. Effect of previous nitrate deprivation on ^{15}N-nitrate absorption and assimilation by wheat seedlings. Journal of Plant Physiology. 1984;**16**:113-122

[43] Gojon A, Soussana JF, Passama L, Robin P. Nitrate reduction in roots and shoots of barley (*Hordeum vulgare* L.) and corn (*Zea mays* L.) seedlings. I.^{15}N study. Plant Physiology. 1986;**82**:254-260

[44] Gojon A, Plassard C, Bussi C. Root/shoot distribution of NO_3^- assimilation in herbaceous and woody species. In: Roy J, Garnier E, editors. A Whole Plant Perspective on Carbon-Nitrogen Interactions. The Hague, The Netherlands: SPB Academic Publishing; 1994. pp. 131-147

[45] Abdel-Latif S, Kawachi T, Fujikake H, Ohtake N, Ohyama T, Sueyoshi K. Contribution of shoots and roots to in vivo nitrate reduction in NADH-specific nitrate reductase-deficient mutant seedlings of barley (*Hordeum vulgare* L.). Soil Science & Plant Nutrition. 2004;**50**:527-535

[46] Jackson WA, Volk RJ. Attributes of the nitrogen uptake systems of maize (*Zea mays* L.): Maximal suppression by exposure to both nitrate and ammonium. NewPhytologist. 1995;**130**:327-335

[47] Aslam M, Huffaker RC. In vivo nitrate reduction in roots and shoots of barley (*Hordeum vulgare* L.) seedlings

in light and darkness. Plant Physiology. 1982;**70**:1009-1013

[48] Oji Y, Otani Y, Hosomi Y, Wakuichi N, Shiga H. Nitrate reduction in root and shoot and exchange of reduced nitrogen between organs in two-row barley seedlings under light-dark cycles. Planta. 1989;**179**:359-366

[49] Delhon P, Gojon A, Tillard P, Passama L. Diurnal regulation of NO_3^- uptake in soybean plants. I. Changes in NO_3^- influx, efflux, and N utilization in the plant during the day/night cycle. Journal of Experimental Botany. 1995;**46**:1585-1594

Section 2

Plant Microbe Interactions

Chapter 4

The Role of Plant Growth-Promoting Bacteria in the Growth of Cereals under Abiotic Stresses

Martino Schillaci, Sneha Gupta, Robert Walker and Ute Roessner

Abstract

Plant growth-promoting rhizobacteria (PGPR) are known to improve plant performance by multiple mechanisms, such as the production of beneficial hormones, the enhancement of plant nutritional status, and the reduction of the stress-related damage. The interaction between plants and PGPR becomes of particular interest in environments that are characterized by suboptimal growing conditions, e.g., high or low temperatures, drought, soil salinity, and nutrient scarcity. The positive role of PGPR will become even more appealing in the future, as world agriculture is facing issues as climate change and soil degradation. This chapter aims to discuss the main mechanisms of the interaction between PGPR and plants and will focus of how PGPR can decrease abiotic stress damage in cereals, which are critical crops for human diet.

Keywords: PGPR bacteria, global warming, abiotic stresses, cereals, growth-promoting mechanisms

1. Introduction

Global agriculture is facing the difficult challenge of increasing the productivity and output required to feed a growing population. Additionally, fertile land areas available for agriculture are gradually decreasing due to climate change, soil degradation, and pressure from urban developments. These concerns are particularly relevant as they negatively affect yields of cereal crops, which are a fundamental diet component in global society [1].

To help overcome this problem, researchers have turned their attention to understanding interactions between plants and soil microorganisms. Plant roots interact with the soil microbiota, which have various effects on plant growth and development, ranging from beneficial to pathogenic [2]. Plant growth-promoting rhizobacteria (PGPR) play important, but still poorly understood, roles in plant growth promotion, especially under environmental stress such as drought, temperature, and salinity [2–4].

There are various mechanisms through which PGPR improve plant performance, often in a synergic manner; some examples include the production of plant growth-promoting hormones, improvement of plant nutritional status, and decreased stress damage [2]. Interactions between plants and PGPR can result in improvement of plant performance and enhanced resistance to biotic and abiotic stresses which are important traits for cultivated crops [5].

2. Importance of cereals in global nutrition

Cereals are annual plants belonging to the monocotyledonous Poaceae family and are a vital food source for humans as they provide almost one half of the calories that are consumed daily in the world [6]. Furthermore, cereals are also extensively used as animal feed, mainly for livestock and poultry, and as raw materials for many industrial processes, primarily the production of alcoholic beverages [1].

In the last 50 years, the increase of cereal production (+240% in the time window 1961–2017 shown in **Figure 1**) is the result of increased yields per hectare (+201%) rather than the expansion of land allocated to cereal production (+12%) (**Figure 1**). However, this trend has recently decreased. The average production rate of cereals was 3.6% per year between 1961 and 2007, and it decreased to an average of 2.7% between 2007 and 2017 [7]. This is likely to be linked to multiple factors, including climate change, soil degradation, use of soil for non-alimentary purposes, restrictions on water, nutrients and land for agriculture, and limitations of traditional breeding.

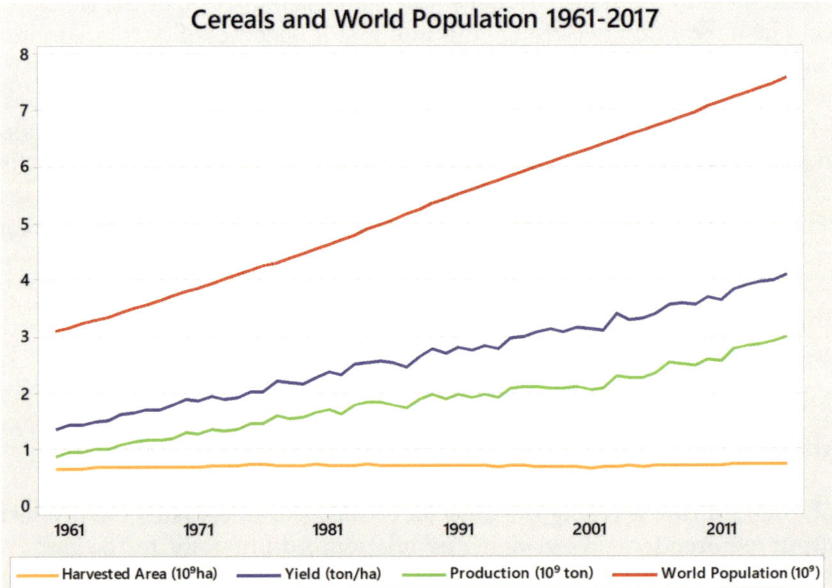

Figure 1.
Cereal cultivation records and world population data since 1961. Cereals cultivated land, soil productivity as yield, world grain production, and world population are displayed [7].

3. Abiotic stress effects on agriculture

Most cultivated soils in the world are characterized as being suboptimal. Any deviation from optimal growth conditions causes several interconnected reactions in plants that can be described as an attempt to adapt to new environmental conditions in an effort to maintain homeostasis. If the stress endures too long or is too severe, it can permanently damage plant physiology or result in death. While many plants are able to adapt to stress, the process requires energy that is diverted from active growth, resulting in smaller acclimated plants [8]. Abiotic stresses, that is, stresses caused by nonliving factors, are thought to be the main cause of global crop loss with decreased productivity of more than 50% annually [9]. Drought and salinity stress are potent environmental hazards for agriculture, particularly in arid and semiarid regions which are already approaching the limits of crop productivity, and due to global warming and degradation of agricultural soils, these regions may no longer support crop plants in the future [10, 11].

3.1 Climate change

Food security is positively correlated with social and economic stability; given climate change is threatening food production, there are extended and complex implications. Since the mid-nineteenth century, average temperatures have increased by 0.8°C, and by the end of this century, temperatures are predicted to increase between 1.8 and 4°C compared to the end of the last century [12]. This change is causally related with human activities by the production of greenhouse gases such as carbon dioxide, the concentration of which rose from ~284 ppm in 1832 to 397 ppm in 2013 [13].

While CO_2 is generally accepted as a greenhouse gas, there is now increased interest in the role of nitrous oxide (N_2O). This compound can originate from the denitrification of N fertilizers, which are commonly used in modern agriculture. In 2014–2015, more than half of all N fertilizer was applied to cereal crops alone [14]. The reintroduction of N in N-depleted soil is an essential agricultural practice that has led to increased yields over the last few decades. However, the application of N fertilizer is inefficient, and it is estimated that only one third of the applied N is absorbed by plants, with the excess being lost in surface runoff, leaching in groundwater, or volatilization into the atmosphere [15]. Atmospheric N_2O, while less abundant than CO_2, is 300 times more potent as a greenhouse gas [16].

Climate change caused by greenhouse gas emissions is predicted to directly impact the productivity of agricultural systems in almost every part of the planet. While many agricultural sites in cold-continental areas will benefit from the increased temperatures, regions characterized by temperate, tropical arid, or subarid climates are likely to face decreasing yields [17]. By modeling the effects of climate change on the yields of various cereals in different areas of the world, it was predicted that by the end of the century, heat stress events will increase in areas of Central and Eastern Asia, Southern Australia, Central North America, and Southeast Brazil (rice); Northern India, the Sahel region, Southeast Africa, and Central South America (maize); and Central Asia (wheat) [18]. Kompas et al. [10] estimated that if no measures are taken to reduce greenhouse gas emissions, the average world temperature increase of 4°C by 2100 will severely decrease food production in almost all countries in the world. This will result in economic loss of approximately 23 thousand billion US$ on average, with Southeast Asia and developing countries of Africa predicted to face the largest losses (21 and 26% of GDP, respectively).

3.2 Agricultural soil degradation

Soil degradation is one of the main concerns impacting agricultural productivity, especially in tropical and subtropical areas [19]. Globally, one third of land is affected by some form of deterioration [20]. Unsuitable agricultural techniques, together with excessive crop residue removal and unbalanced use of chemical fertilizers, can decrease soil quality, deplete organic matter stocks, and increase erosion. Crop removal from the production site causes the loss of elements that are essential for plant growth, and these elements must be constantly reintroduced to avoid productivity decreases [21].

Using soils for agricultural purposes can cause degradation of water sources, due to leaching of degraded fertilizers into groundwater. Many rivers in developing countries have severe water pollution and eutrophication issues. Irrigation is an essential management strategy to obtain sufficient productivity to meet food demands in many arid and semiarid areas, but it can lead to undesirable effects. Improper irrigation techniques have increased saline-sodic soils that now occur in more than 20% of irrigated lands [22].

4. Plant growth-promoting bacteria

A common misconception during the nineteenth century was that healthy plants should be sterile, not interacting with any microorganisms. This assumption was initially questioned by Victor Galippe [23], who proved that healthy plants could host various microbes in their tissues. Today, we know that almost all terrestrial plants from various environments interact with the surrounding microbiota during all stages of plant development. The relationship between host plant and microbe can range from parasitism, commensalism or mutualism, or neutral or beneficial for plant growth and can vary greatly due to a multitude of factors, both biotic and abiotic. PGPR are attracted to plants by organic exudates released through roots and colonize the root surface and the soil directly in contact with the root. The soil matrix directly in contact with plant roots is called the rhizosphere [24], and the extracellular surface of roots is termed the rhizoplane [25]. Here, colonizing microorganisms can establish the exchange of nutrients and various compounds with the plant, summarized in **Figure 2**.

Nutrients and organic compounds released into the rhizosphere from roots are derived from photosynthesis, and plants release up to 30% of their photosynthates through the roots [26]. These include a variety of compound classes such as carbohydrates, amino acids, organic acids, flavonoids, and lipids that can be used as energy sources for microbes [27]. The sensing and active migration of bulk soil bacteria toward these compounds is called chemotaxis, leading bacteria to colonize the rhizosphere and rhizoplane [28]. By producing exudates, plants can select bacterial species that are attracted to specific compounds, thereby directing the abundance and diversity of microbes in the rhizosphere [29, 30]. Wild oat has been reported to modify the bacterial population of its rhizosphere enriching mainly the Firmicutes, Actinobacteria, and Proteobacteria [31]. The latter group in particular is commonly believed to be the main microbial component in PGPR interactions, due to their capacity for fast growth and diverse metabolic pathways capable of utilizing a great variety of exudate compounds as an energy source [29]. In the model cereal plant *Brachypodium*, the rhizosphere microbiome changes not only within the loosely bound rhizosphere soil and tightly bound rhizosphere soil but also within seminal and nodal roots [32]. It is noteworthy that plants can indirectly influence the colonization of the rhizosphere, by changing the environment conditions. Some examples are changes in pH levels by ion uptake, the reduction of O_2 and H_2O levels caused by root respiration, and water absorption [29].

Different types of root exudates can attract different PGPR. For example, various strains of *Azospirillum brasilense*, a gram-negative Alphaproteobacteria, showed different degree of attractions to various compounds released by different host plants [33]. The composition of root exudates can vary greatly among different plant species. Two different studies [34, 35] reported how even different genotypes of the same plant species can host different bacterial populations in their rhizosphere. Exudates vary between different parts of the roots, different developmental stages of the plant, or as a response to different growth conditions [36]. This means that the same plant can interact with a multitude of different soil bacterial strains over time and space [37].

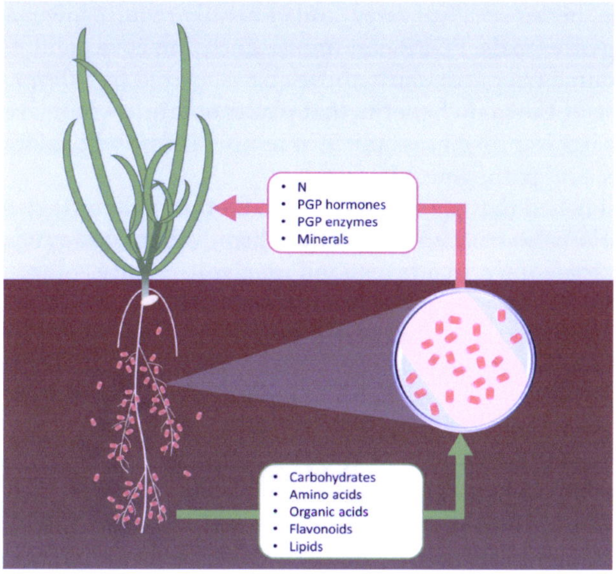

Figure 2.
A model of interactions between plants and PGPR. Exudates released by plant roots attract soil bacteria that can colonize rhizosphere and/or plant tissues. Here, they provide various beneficial compounds to the plant in exchange of nutrients, mainly photosynthates.

Nehl et al [38] use the term "rhizobacteria" to describe rhizoplane/rhizosphere bacteria, but there are also endophytic bacteria that can reside inside plant tissues. To date, numerous interactions between plants and rhizosphere-/rhizoplane-colonizing bacteria have been described, but some microbes are even more specialized. Once they have colonized the rhizoplane, they are able to penetrate root tissues and directly access apoplastic organic compounds, thereby avoiding competition with other microbes in the rhizosphere [39]. Root penetration can be both active, by the production of cell wall-degrading enzymes such as cellulase, and passive, for example, entering via the cracks that form on the root surface during lateral root development [40]. Colonization beyond the rhizosphere into the apoplast requires specialized microbial morphology. Czaban et al. [41] described how the occurrence of flagellar motility in bacterial strains isolated from the internal root tissue of wheat was five times higher than what was observed in bacteria isolated from the rhizosphere.

Bacillus, *Pseudomonas*, *Enterobacter*, *Klebsiella*, *Serratia*, and *Streptomyces* are some of the most commonly found genera of endophytic bacteria in plant tissues [42]. By passing the endodermis, many bacterial species are able to spread from the roots, reaching and colonizing other organs of the stems [43]. Endophytic bacteria can also spread from plant tissue to seeds becoming the starting inoculum for the colonization of subsequent generations of plants. The transmission of bacteria through generations of plants is a process known as vertical transmission. Johnston-Monje and Raizada [44] described how modern varieties of maize and their wild ancestors share common endophytic bacteria communities hosted in their seeds, and a following study conducted on wheat demonstrated how these communities play a positive role in plant growth [45].

4.1 Plant growth promotion driven by rhizobacteria

Galippe's intuition that plants interact with microbes throughout their life led to a significant increase in the comprehension of the beneficial role that bacteria can have on plant growth. PGPR interactions can result in higher plant biomass, higher nutritional value, better survival rates, and generally require lower agricultural inputs. Focusing on cereals, PGPR can significantly improve plant performance in several environments, particularly those characterized by suboptimal growth conditions. Some of the main benefits that plants obtain are increased root development which imparts improved resistance to temperature and osmotic stress, soil pollutants, pests, and pathogens [46].

It is well established that plant responses to biotic and abiotic stresses require complex adaptations to structure and metabolism. When biotic and abiotic stresses are applied simultaneously, plants respond much differently compared to stresses applied separately [9]. It is therefore reasonable to assume if a plant is exposed to both biotic and abiotic stresses that PGPR may directly mitigate the effect of biotic stresses by improving plant resistance to abiotic stresses.

4.1.1 Hormone-related mechanisms

The most well-described mechanism by which PGPR can improve cereal productivity is the productions of various plant growth-promoting hormones that usually co-affect the performance of the plant in a highly integrated manner [47]. Auxins are a class of hormones typically synthesized by apical buds, and from there they are transported to other parts of the plant. In this class of hormones, the most characterized is indole-3-acetic acid (IAA), which enhances cell elongation and differentiation and, in roots, stimulates lateral root development [42, 48]. Various reports have shown how the production of auxins from PGPR is one of the most important mechanisms for plant growth promotion. Barbieri and Galli [49] inoculated wheat with two strains of *Azospirillum brasilense*, of which one was a mutant with impaired IAA production. They observed how only the wild-type strain promoted lateral root development, a result that suggests a primary role of IAA in improving plant root development. IAA can indirectly improve the nutritional status of the plant by increasing root development (specifically lateral roots), hence allowing the plant to explore a higher portion of soil substrate, an important trait particular for the acquisition of low-mobility nutrients such as phosphorus [50].

Gibberellins (GAs) can be produced by PGPR [51] and are believed to play an important role in promoting plant growth. These diterpene hormones are naturally present in plants, regulating key processes such as seed germination, stem elongation, leaf expansion, root growth, and root hair abundance [52, 53]. One of the best known GAs is GA_3, commonly known as gibberellic acid, which plays a key role in determining plant source-sink relations. The role of gibberellins in the response of cereals to stresses varies depending on the stress type [54], but in general, plants tend to reduce GAs levels when growing in suboptimal conditions. The exogenous application of gibberellins has been reported to improve wheat and rice performance undergoing saline stress [55, 56] and to reduce heavy metal stress symptoms in rice [57].

Many PGPR are able to degrade 1-aminocyclopropane-1-carboxylic acid (ACC) through the enzyme ACC deaminase and use the degradation products as a nitrogen source [42]. ACC is the biosynthetic precursor of ethylene, a hormone naturally present in plants, and its abundance is often increased in response to stresses. While at optimal levels, ethylene is involved in essential processes such as tissue differentiation, root development, flowering, grain development, and natural tissue senescence and abscission; when overproduced it can decrease plant performance [58]. In abiotically stressed plants, the increase of ethylene can trigger chlorosis and early maturation and senescence of organs, seeds in particular [59, 60], and have an inhibitory effect on root growth [42]. By impairing the ethylene signaling pathway, the interaction with PGPR can decrease the stress-related damage in the plant [2].

Similar to ethylene, abscisic acid (ABA) is a hormone commonly produced by plants in response to various types of stress, particularly osmotic stress [61]. Naturally involved in seeds and buds dormancy, ABA shares the first biosynthetic steps with cytokinins, a phytohormone class that often plays an antagonistic role to ABA. In dry or saline soils, reactive oxygen species (ROS) increase the biosynthesis of ABA, which is then transported to leaves, where it causes stomatal closure to reduce transpiration and water loss [62]. As a consequence, the diffusion of CO_2 into leaves is decreased, lowering photosynthetic rates [63, 64]. PGPR have been reported to increase the resistance of plants to salinity, hence decreasing the stress-related ABA accumulation in plants and preserving photosynthetic efficiency [65, 66].

5. Plant-bacterial interactions enhance abiotic stress responses

Bacteria can have various effects on their host plant. PGPR can affect plant growth both directly, such as by fixing atmospheric N_2 into biologically available N compounds or by producing growth-promoting hormones [52], and indirectly, by preventing the growth of plant pathogens or increasing plant resistance to them [43]. A necessary condition for bacteria to be beneficial to a plant is rhizosphere competence as the competition and conditions in the rhizosphere are vastly different to that of bulk soil. The rhizosphere contains a higher abundance of bacteria than bulk soil, but the diversity is much lower. The colonization of the root system of plants is not homogenous; the density of specific bacteria varies in different parts of the root system and is likely to be related to different root exudates released by different parts of the roots [37]. Another mechanism likely to regulate the colonization of the rhizosphere is bacterial quorum sensing, which is the regulation of gene expression driven by bacterial population density and can occur both within bacteria of the same species and among different species [67]. Quorum sensing can influence the bacterial competitiveness, therefore affecting the roots colonization patterns [37].

5.1 Thermic stress adaptation

Temperature stress causes a shift in hormone production, particularly ethylene, which can often impair plant growth [58]. High-temperature stress causes denaturation and aggregation of cellular proteins that, if left unchecked, leads to cell necrosis. Imbalance between ABA and cytokinins derived from prolonged heat stress during the reproductive stage can lead to grain abortion [68]. Heat responses include inhibition of normal transcription and translation and increased expression of genes coding for heat shock proteins and thermotolerance induction [69]. Low-temperature stress, conversely, damages metabolic processes, changes membrane properties, causes structural changes in proteins, and inhibits enzymatic reactions [70]. If it occurs during spore formation, cold can cause sterility of flowers by interfering with meiosis [71].

The literature on PGPR interactions with cereals at suboptimal temperatures is relatively scarce, and the mechanisms by which cereals adapt are not well defined. It is suggested that the geographical origin of the bacteria determines the optimal growth range at which they interact beneficially with plants. In a study on wheat, bacteria isolated from cold climates have been reported to efficiently colonize the plant rhizosphere and improve their resistance to low-temperature stress, and the same trend was observed when wheat plants were inoculated with bacteria isolated from warm environments and subjected to high-temperature stress [72]. It is possible that the bacteria isolated from different temperatures can outcompete the indigenous microbial population by tolerating either cold or warm conditions giving rise to a higher abundance and colonization of the rhizosphere.

Inoculation with a *Pseudomonas aeruginosa* strain isolated from a hot semiarid environment improved survival rate, development, and biochemical parameters of sorghum seedlings when the plants were exposed to heat treatment, while the biomass production was not affected at optimal temperatures [73]. In another study, various cold-tolerating *Pseudomonas* spp. were inoculated onto wheat grown at low temperatures, giving analogous results. The authors suggest the beneficial effect was linked to a better root development in inoculated strains that improved nutrient uptake and, in general, caused a better adaptation to cold [74].

As global warming threatens to change significantly the temperature of most cultivated lands [17], the development of cereals with enhanced adaptation capacity to heat or cold stress is an essential task in order to sustain profitability and production at suboptimal temperature conditions. While further research is necessary to better understand the mechanisms that regulate PGPR-plant interactions in such conditions, the studies done so far suggest how PGPR can be a valuable source of temperature-stress resistance, especially when they evolved in areas characterized by warm or cold climates, depending on the case.

5.2 Osmotic stress adaptation

Both dry and saline soils can cause osmotic stress in plants, which results in cell dehydration due to lack of water (drought) or unavailability of water (salinity). These two stresses are often agronomically significant, as high salinity in soil is mainly caused by irrigation, a necessary practice for increasing yields in many areas of the world characterized by insufficient rainfalls. When water supply is insufficient to remove ions from superficial soil layer, they accumulate causing an increase of salinity [75].

Salinity is also the result of land clearing, as deep subsurface roots no longer are able to keep the water table below ground level. As the water table rises, it brings with it saline water that can render hundreds of square kilometers of agricultural land uncultivable [76]. Plants growing on such soils often suffer from osmotic stress that reduces water absorption and increases ionic concentration in tissues to toxic levels [77]. PGPR can decrease these stress symptoms through various mechanisms, such as production of Na^+-binding exopolysaccharides [78], improvement of ion homeostasis [79], decrease of ethylene levels in plants through ACC deaminase [80], and synthesis of IAA [81]. Wheat seeds inoculated with a species from the genus *Pseudomonas* showed increased germination rates in a saline environment; Egamberdiyeva [82] ascribed this to the production of plant growth regulators by the bacteria.

Drought is considered as the major cause of yield loss [83], negatively affecting most physiological processes in plants. Plant cells respond to water loss by increasing the production of abscisic acid (ABA) in roots that increases water uptake and causes leaf stomatal closure and reduces leaf expansion to reduce dehydration [84]. Smaller leaves cause impaired photosynthesis, consequently decreasing dry matter accumulation and grain yield [85]. Under water deficiency, both cell division and enlargement are lowered due to damaged enzyme activities, leading to overall smaller plant organs. Grain production is also reduced in cereals due to flower abortion [86, 87].

Plants often react to drought by increasing the amount of osmolytes in their tissues and consequently increase their osmotic potential [88]. Drought can also cause an increase of ROS in plant tissues. Proline, an amino acid whose abundance is increased under water deficiency, can both work as an osmolyte and scavenger for ROS under stress [89]. In general, PGPR can improve the performance of plants in dry environments by exudating osmolytes that increase the osmotic potential of plants [42, 90, 91].

Another mechanism for improving resistance to drought is the synthesis of beneficial hormones (IAA) and enzymes (ACC deaminases) and the decrease of stress-related hormones such as ethylene and ABA in the plant. Naveed et al. [92] reported that two maize cultivars exposed to drought showed reduced damage when inoculated with two different PGPR, probably due to hormones produced by the bacteria and stress-reducing enzymes synthesized by both the plants and the bacteria during the interaction. Wheat plants inoculated with various PGPR showed an improved resistance to salt and drought treatments, linked to decreased ABA and ACC levels in plant tissues [65]. In a similar study [66], rice plants showed decreased endogenous ABA levels and increased biomass when inoculated with *Bacillus amyloliquefaciens*; the authors hypothesize that inoculation increased salt tolerance in plants through an ABA-independent pathway, and this prevented the stress-dependent ABA accumulation and the resulting growth impairment [63].

Sarig et al. [93] report that sorghum plants subjected to osmotic stress after their emergence showed decreased damage when colonized by *Azospirillum brasilense*. It is unclear, however, if the observations were a drought-specific response or an indirect effect of inoculated plants showing a better root development and higher hydraulic conductivity at the time of the stress. In two successive studies [94, 95] conducted on various *Azospirillum* spp., inoculated wheat plants subjected to drought had decreased grain loss, better water status, and higher K and Ca content, with the latter in particular suggested to be involved in the adaptation of the plants to environmental stress. Bacterial nitrate reductase was also suggested to play an important role in nitrate assimilation of plants under drought [95].

As previously mentioned, drought and saline stress are related, since salinity is often the result of irrigation practices to avoid plant desiccation from drought stress. This concern may become more relevant in future years, as higher temperatures caused by global warming will result in higher evapotranspiration, hence requiring increased irrigation. By the year 2050, 50% of all arable lands might be affected by serious salinization [96]. Improving the resistance of plants to dry environments would decrease the necessity of irrigation, indirectly decreasing the ongoing salinization process in agricultural land.

5.3 Improvement of the plant nutritional status

In natural environments, plants die and decompose where they grew, and the subsequent detritus reintroduces soils with most of the nutrients they absorbed during their growth. In cultivated lands, those nutrients are removed at harvest and must be constantly replaced to avoid productivity decrease. Among the macronutrients, nitrogen, phosphorus, and potassium are the most important for plant growth, and they are typically reintroduced using synthetic fertilizers. Unbalanced use of fertilizers can decrease soil quality, consume organic matter stocks, and increase erosion risk. Soil bacteria can improve the nutritional status of plants directly by increasing nutrient bioavailability and/or indirectly by improving plant root development, hence allowing them to explore higher areas of soil [97].

5.3.1 N_2 fixation and absorption

Several bacterial species are classified as diazotrophs, which are microorganisms that are able to utilize the nitrogenase enzyme to fix atmospheric N_2. Diazotrophic bacteria can fix N_2 in either a free-living form or in association with a host as an endosymbiont. The most well-described interaction between plants and diazotrophic bacteria is the rhizobia-legume symbiosis. Rhizobia are a group of various Proteobacteria that can colonize plant roots and fix atmospheric nitrogen, which is then partly provided to the plant in exchange of photosynthates [98]. While this association has been observed mainly in legumes, some species of rhizobia can also colonize cereals. Gutierrez-Zamora and Martinez-Romero [99] showed how maize and bean plants cultivated in association shared the same *Rhizobium etli* strains, with the bean plants probably constituting the source of inoculum for maize. The interaction with the rhizobia increased the biomass of both crops, but in maize this outcome might have been linked to mechanisms other than N_2 fixation, such as hormone production. Rice inoculated with an *Azoarcus* sp. showed improved growth regardless of colonization by the wild-type strain or with a mutant strain deficient in the nitrogenase genes [3]. When spring wheat and maize were inoculated with two different rhizobia and grown at various soil N levels, the two strains were effective in enhancing plant growth only at low and intermediate levels of soil N. The authors suggest that plant growth-promoting hormones released by the bacteria caused a better root development in inoculated plants that were able to absorb more nutrients from the soil [100].

In general, diazotrophic bacteria associated with cereal roots often carry the nitrogenase genes necessary for the fixation of atmospheric nitrogen, but the relative enzymes are not always synthesized inside plant tissues. Furthermore, the amount of fixed N provided to the plant is often negligible, due to low presence of diazotroph bacteria or because bacteria use fixed nitrogen for their own growth [101]. The nitrogenase enzyme cannot function in the presence of O_2, so it may be desirable to engineer free-living diazotrophic bacteria that are able to colonize plant tissues. Other possible ways might be to increase the fixing bacteria population by engineering plants capable of exudating diazotroph favorable compounds or engineering bacteria capable of providing the plant with higher levels of nitrogen [102].

Fox et al. [103] modified a *Pseudomonas* sp. genome by adding a gene cluster with nitrogenase activity that improved the performance of wheat and maize by fixing N_2. This is an example of some of the approaches toward nitrogen-fixing cereals, that is, plants capable of sourcing the N necessary for their growth from the atmosphere via endosymbionts [104].

Farmers have benefited from the rhizobia-legume symbiosis for centuries, and extending this characteristic to cereals would be a decisive benefit for modern agriculture, providing a continuous, ecologically, and economically sustainable source of N to the most important crops.

5.3.2 Improvement of soil nutrient uptake

Despite the benefits PGPR impart on plant nutrient content, it is often unclear if this improvement is related to an enhanced mineral uptake or if it is the result of improved root system development in inoculated plants due to bacterial hormones and/or enzymes [48].

Various bacterial strains are known to increase bioavailability of phosphorus in soil, due to the mineralization of organic phosphate and solubilization of inorganic phosphate. Some of the bacterial compounds linked to these two processes are acid phosphatases and organic acids, respectively [105]. Phosphate-solubilizing bacteria have been reported to improve the growth of maize [106], rice [107], and wheat [108].

PGPR can also synthesize siderophores that are low-molecular-weight compounds with high iron-binding affinity [109] that can complex with Fe (predominantly Fe^{3+}) in soil. The iron-siderophore complex is then assimilated by the bacterium using a complex-specific receptor [110]. This has various effects, it depletes the soil iron supply, thereby preventing the growth of other potentially pathogenic microbes, and, if the iron is then provided to the plant, it can directly improve plant growth [48]. Furthermore, the bacterial nitrogenase activity and *nif* gene expression are iron dependent [111, 112]; hence, the absorption of iron from the soil enables diazotroph bacteria to convert atmospheric N_2 to a form that is bioavailable for the plant.

PGPR can indirectly improve plant performance neutralizing the stress-related hormones produced by the plant in poor soils. Wheat plants grown at various levels of N, P, and K, showed increased grain yield and biomass production when colonized by *Pseudomonas* spp., with the bacterial growth promotion being negatively correlated with the amount of provided nutrients [97]. The authors ascribe this outcome to bacterial production of ACC deaminase that decreased ethylene levels produced by plants as a response to low nutrients levels, which impaired root development in uninoculated plants.

Overall, plant growth promotion is ascribed to a combination of multiple mechanisms. Egamberdiyeva [113] inoculated maize seeds with PGPR with nitrogenase and/or IAA activity and grew them on two soil types with different nutrient availabilities. Inoculated plants generally developed a higher root and shoot biomass and had higher N, P, and K contents, the improvement being more pronounced in plants grown on nutrient-poor soils. However, this study did not consider the possible interactions of inoculated strains with the native microbial populations that may have affected the results.

In 2014–2015, out of 182 million metric tons (Mt) of consumed fertilizer, one half was applied to cereals [14]. Cereals consumed more than one half of N fertilizers and more than one third of P and K fertilizers. As previously mentioned, these amendments have both a high economical and environmental cost, as they can cause soil degradation, pollution of water, and eutrophication. While developing N-fixing PGPR is a task yet to be achieved in cereal agriculture, it is well documented how PGPR can improve the efficiency of nutrient uptake in crops. This can occur by either increasing the bioavailability of nutrients in the soil or as a consequence of better root development, resulting in better soil exploration.

6. Issues and perspectives

Cereal-PGPR interactions have been widely studied over the last few decades, and the positive influence that they can have on plant growth is still being established. However, the lack of consistency among different studies is still a concern, highlighting that when multiple biological actors are involved, no generalizations can be made. The same bacterial strain can be beneficial to a plant species and damage another [114] or have no effects or even be detrimental for plant performance when the growing conditions are optimal but become beneficial when growing conditions worsen [2–4]. In two studies on maize and rice subjected to water deficiency [90, 115], the beneficial effects of various bacterial isolates on plant growth increased with the severity of the stress. Studying the interaction between PGPR and gum rockrose (*Cistus ladanifer*), Solano et al. [116] hypothesized that a possible explanation for this is that poor environments may impair the growth of indigenous microbial communities, this way decreasing the competition for those microbes that establish advantageous relationships with plants. Another possible explanation is that when the main bacterial mechanism of plant growth promotion is providing them with nutrients, the benefit might be limited in nutrient-rich soil, while it can be significant in the case of limiting nutrients [117].

The observed outcomes change particularly from laboratory and climate chamber trials to more open setups such as greenhouses and field, in which bacteria often fail to improve plant growth [37]. Most of the studies conducted so far on the interaction between cereals and PGPR were performed in controlled environments, usually applying only one single stress at a time. While this is a necessary compromise when starting to study this interplay, it often entails a significant bias from realistic field environment [2], in which plants frequently face more variable growing conditions and face multiple stresses at the same time, triggering unique responses in plants that are different from the sum of plant response to stress applied individually [118]. So far, very few experiments have studied the interaction between bacteria and crops under multiple stresses, but replicating as accurately as possible real field conditions is an essential step for understanding and exploiting the role of PGPR in agriculture. In addition to the more unstable growing environment, another important variable added in field experiments is the interaction with the native microbiota. Often inoculated bacteria in the field show lower rhizosphere or root colonization than laboratory, climate chamber, and greenhouse trials [119], in which the growth medium is usually sterilized at the beginning of the experiment.

One of the hypotheses that can be drawn from the current literature is that the origin of the inoculated bacteria is often a decisive factor for the interaction to improve plant growth. Bacteria isolated from the same plant species used in trials are more likely to play a beneficial role, probably due to the plant-specific exudates that have a key role in the early phases of the interaction [100]. Similarly, bacteria isolated from environments characterized as suboptimal (temperature in particular) that are similar to the conditions and stress applied in plant trials may be more beneficial than bacteria isolated from optimal conditions, delivering more benefits to the plant, due to adaptations that allow the bacteria to be more competitive than the native microbiota [72]. Unfortunately, inoculum used in trials may become less effective due to continual cultivation in laboratory environments, and when planning a plant trial, this should be taken into consideration.

One of the problems facing commercialization of PGPR on markets is the inoculation delivery method on plants. In the laboratory, a common method is dip inoculation where seedling roots are immersed in bacterial culture and then transplanted into the growth substrate, but this approach is not feasible for annual cereals on the field scale. The on-field application of bacterial solutions after seedling germination, while less laborious, still requires considerable equipment and technical knowledge. The most feasible way to apply PGPR on field is probably the use of pre-inoculated seeds (this is already used for rhizobia-legume inoculation) allowing farmers to bulk sow, relieving them from the inoculation step. When the seed bacterial treatment is done immediately before germination, the required strength of bacterial inoculum is typically smaller than in seedling treatments, but ideally inoculants should survive long enough on seed coats to be present during germination; however, prolonged survival of microbial treatment on seeds is still a challenge [120]. Moreover, inconsistencies between performances of seed inoculants are often observed in different trials, and further research is required to address this issue [121]. Utilizing vertical transfer of microbial endosymbionts in seeds may also present a possible inoculation technology that has not been explored extensively and may provide economic benefits to farmers [120] and could potentially mitigate the problem of inoculum viability in seed coats. Recently, studies on bacterial strains vertically transmitted in cereal seeds have shown promising plant growth-promoting effects, likely linked to their ability to solubilize phosphorus, produce hormones, siderophores, and ACC deaminase [122]. By exploiting the existing interactions between plants and known seed endophytic bacteria or isolating new bacterial strains capable of inhabiting seeds for vertical transmission by crops, new technologies may emerge that have large-scale economical applications.

During the last decades, selection of crops has been driven by increased productivity in nutrient-rich environments, with scarce focus on the positive effects of PGPR, and this trend might have led to the loss of plant traits associated with the microbial interaction [5]. The identification and reintroduction of the genes associated with those traits might enhance the positive effects of PGPR, especially in poor environments, and selecting plants that have superior interaction with rhizosphere microbiota should be considered in plant breeding programs. Additionally, a more immediate way to alleviate temperature stress could be to inoculate plants with bacteria originated from hot-climate regions that as a consequence are more likely to help their host to perform better in a warming environment [29, 72].

The interaction with microbes will gain more attention in the future, considering the effects of climate change, due to the microbial genetic plasticity compared to plants. PGPR may evolve rapidly, developing efficient adaptation strategies to the benefit of the plant host as well.

Author details

Martino Schillaci*, Sneha Gupta, Robert Walker and Ute Roessner
School of BioSciences, University of Melbourne, Parkville, VIC, Australia

*Address all correspondence to: mschillaci@student.unimelb.edu.au

References

[1] Sarwar MH, Sarwar MF, Sarwar M, Qadri NA, Moghal S. The importance of cereals (Poaceae: Gramineae) nutrition in human health: A review. Journal of Cereals and Oilseeds. 2013;**4**(3):32-35

[2] Hardoim PR, Van Overbeek LS, Berg G, Pirttilä AM, Compant S, Campisano A, et al. The hidden world within plants: Ecological and evolutionary considerations for defining functioning of microbial endophytes. Microbiology and Molecular Biology Reviews. 2015;**79**(3):293-320

[3] Hurek T, Reinhold-Hurek B, Van Montagu M, Kellenberger E. Root colonization and systemic spreading of *Azoarcus* sp. strain BH72 in grasses. Journal of Bacteriology. 1994;**176**(7):1913-1923

[4] Lynch JM, de Leij F. Rhizosphere. e LS; 2001

[5] Germida J, Siciliano S. Taxonomic diversity of bacteria associated with the roots of modern, recent and ancient wheat cultivars. Biology and Fertility of Soils. 2001;**33**(5):410-415

[6] FAO. World Food and Agriculture—Statistical Pocketbook 20182018. p. 254

[7] FAO. FAOSTAT Database [Internet]. 2019. Available from: http://www.fao.org/faostat/en/#data/QC [Accessed: February 25, 2019]

[8] Kosová K, Vítámvás P, Prášil IT, Renaut J. Plant proteome changes under abiotic stress—Contribution of proteomics studies to understanding plant stress response. Journal of Proteomics. 2011;**74**(8):1301-1322

[9] Atkinson NJ, Urwin PE. The interaction of plant biotic and abiotic stresses: From genes to the field. Journal of Experimental Botany. 2012;**63**(10):3523-3543

[10] Kompas T, Pham VH, Che TN. The effects of climate change on GDP by country and the global economic gains from complying with the Paris climate accord. Earth's Future. 2018;**6**(8):1153-1173

[11] Tuteja N. Abscisic acid and abiotic stress signaling. Plant Signaling & Behavior. 2007;**2**(3):135-138

[12] Solomon S, Qin D, Manning M, Averyt K, Marquis M. Climate Change 2007: The Physical Science Basis: Working Group I Contribution to the Fourth Assessment Report of the IPCC. Cambridge University Press; 2007

[13] Tans P, Keeling R. Trends in Atmospheric Carbon Dioxide: Daily Mean Concentration of Carbon Dioxide for Mauna Loa, Hawaii. National Oceanic and Atmospheric Administration, Earth System Research Laboratory (NOAA/ESRL); 2013

[14] Heffer PGA, Roberts T. Assessment of Fertilizer Use by Crop at the Global Level. Paris: International Fertilizer Industry Association; 2017

[15] Garnett T, Conn V, Kaiser BN. Root based approaches to improving nitrogen use efficiency in plants. Plant, Cell & Environment. 2009;**32**(9):1272-1283

[16] Ward M, Jones R, Brender J, de Kok T, Weyer P, Nolan B, et al. Drinking water nitrate and human health: An updated review. International Journal of Environmental Research and Public Health. 2018;**15**(7):1557

[17] Wheeler T, Von Braun J. Climate change impacts on global food security. Science. 2013;**341**(6145):508-513

[18] Teixeira EI, Fischer G, Van Velthuizen H, Walter C, Ewert F. Global hot-spots of heat stress on agricultural crops due to climate change.

Agricultural and Forest Meteorology. 2013;**170**:206-215

[19] Lamb D, Erskine PD, Parrotta JA. Restoration of degraded tropical forest landscapes. Science. 2005;**310**(5754):1628-1632

[20] Bini C. Soil: A Precious Natural Resource. In: Kudrow NJ, editor. Environmental Science and Engineering Technology. Hauppauge, NY: Nova Science Publishers; 2009. p. 1-48

[21] Amundson R, Berhe AA, Hopmans JW, Olson C, Sztein AE, Sparks DL. Soil and human security in the 21st century. Science. 2015;**348**(6235):1261071

[22] Qadir M, Oster J. Crop and irrigation management strategies for saline-sodic soils and waters aimed at environmentally sustainable agriculture. The Science of the Total Environment. 2004;**323**(1-3):1-19

[23] Galippe V. Note sur la présence de micro-organismes dans les tissus végétaux. Comptes Rendus des Séances de la Société de Biologie et de ses Filiales. 1887;**39**:410-416

[24] Smalla K, Sessitsch A, Hartmann A. The Rhizosphere: 'Soil Compartment Influenced by the Root'. Oxford, UK: Blackwell Publishing Ltd; 2006

[25] Foster R. The ultrastructure of the rhizoplane and rhizosphere. Annual Review of Phytopathology. 1986;**24**(1):211-234

[26] Lynch J, Whipps J. Substrate flow in the rhizosphere. Plant and Soil. 1990;**129**(1):1-10

[27] Walker TS, Bais HP, Grotewold E, Vivanco JM. Root exudation and rhizosphere biology. Plant Physiology. 2003;**132**(1):44-51

[28] Lugtenberg BJ, Dekkers LC. What makes pseudomonas bacteria rhizosphere competent? Environmental Microbiology. 1999;**1**(1):9-13

[29] Philippot L, Raaijmakers JM, Lemanceau P, Van Der Putten WH. Going back to the roots: The microbial ecology of the rhizosphere. Nature Reviews. Microbiology. 2013;**11**(11):789

[30] Huang AC, Jiang T, Liu Y-X, Bai Y-C, Reed J, Qu B, et al. A specialized metabolic network selectively modulates *Arabidopsis* root microbiota. Science. 2019;**364**(6440):eaau6389

[31] DeAngelis KM, Brodie EL, DeSantis TZ, Andersen GL, Lindow SE, Firestone MK. Selective progressive response of soil microbial community to wild oat roots. The ISME Journal. 2009;**3**(2):168

[32] Kawasaki A, Donn S, Ryan PR, Mathesius U, Devilla R, Jones A, et al. Microbiome and exudates of the root and rhizosphere of Brachypodium distachyon, a model for wheat. PLoS One. 2016;**11**(10):e0164533

[33] Reinhold B, Hurek T, Fendrik I. Strain-specific chemotaxis of *Azospirillum* spp. Journal of Bacteriology. 1985;**162**(1):190-195

[34] Bouffaud ML, Kyselková M, Gouesnard B, Grundmann G, Muller D, Moënne-Loccoz Y. Is diversification history of maize influencing selection of soil bacteria by roots? Molecular Ecology. 2012;**21**(1):195-206

[35] Peiffer JA, Spor A, Koren O, Jin Z, Tringe SG, Dangl JL, et al. Diversity and heritability of the maize rhizosphere microbiome under field conditions. Proceedings of the National Academy of Sciences. 2013;**110**(16):6548-6553

[36] el Zahar Haichar F, Marol C, Berge O, Rangel-Castro JI, Prosser JI, Balesdent JM, et al. Plant host habitat and root exudates shape soil bacterial community structure. The ISME Journal. 2008;**2**(12):1221

[37] Compant S, Clément C, Sessitsch A. Plant growth-promoting bacteria in the rhizo- and endosphere of plants: Their role, colonization, mechanisms involved and prospects for utilization. Soil Biology and Biochemistry. 2010;**42**(5):669-678

[38] Nehl D, Allen S, Brown J. Deleterious rhizosphere bacteria: An integrating perspective. Applied Soil Ecology. 1997;**5**(1):1-20

[39] Hallmann J, Quadt-Hallmann A, Mahaffee W, Kloepper J. Bacterial endophytes in agricultural crops. Canadian Journal of Microbiology. 1997;**43**(10):895-914

[40] Hardoim PR, van Overbeek LS, van Elsas JD. Properties of bacterial endophytes and their proposed role in plant growth. Trends in Microbiology. 2008;**16**(10):463-471

[41] Czaban J, Gajda A, Wróblewska B. The motility of bacteria from rhizosphere and different zones of winter wheat roots. Polish Journal of Environmental Studies. 2007;**16**(2):301

[42] Dimkpa C, Weinand T, Asch F. Plant-rhizobacteria interactions alleviate abiotic stress conditions. Plant, Cell & Environment. 2009;**32**(12):1682-1694

[43] Compant S, Duffy B, Nowak J, Clément C, Barka EA. Use of plant growth-promoting bacteria for biocontrol of plant diseases: Principles, mechanisms of action, and future prospects. Applied and Environmental Microbiology. 2005;**71**(9):4951-4959

[44] Johnston-Monje D, Raizada MN. Conservation and diversity of seed associated endophytes in Zea across boundaries of evolution, ethnography and ecology. PLoS One. 2011;**6**(6):e20396

[45] Herrera SD, Grossi C, Zawoznik M, Groppa MD. Wheat seeds harbour bacterial endophytes with potential as plant growth promoters and biocontrol agents of Fusarium graminearum. Microbiological Research. 2016;**186**:37-43

[46] Lugtenberg B, Kamilova F. Plant-growth-promoting rhizobacteria. Annual Review of Microbiology. 2009;**63**:541-556

[47] Trewavas A. Signal perception and transduction. In: Buchannan B GW, Jones R, editor. Biochemistry and Molecular Biology of Plants. USA. 2000. pp. 930-987

[48] Glick BR. The enhancement of plant growth by free-living bacteria. Canadian Journal of Microbiology. 1995;**41**(2):109-117

[49] Barbieri P, Galli E. Effect on wheat root development of inoculation with an *Azospirillum-brasilense* mutant with altered indole-3-acetic-acid production. Research in Microbiology. 1993;**144**(1):69-75

[50] Wittenmayer L, Merbach W. Plant responses to drought and phosphorus deficiency: Contribution of phytohormones in root-related processes. Journal of Plant Nutrition and Soil Science. 2005;**168**(4):531-540

[51] Bastian F, Cohen A, Piccoli P, Luna V, Baraldi R, Bottini R. Production of indole-3-acetic acid and gibberellins a(1) and a(3) by acetobacter diazotrophicus and herbaspirillum seropedicae in chemically-defined culture media. Plant Growth Regulation. 1998;**24**(1):7-11

[52] Bottini R, Cassan F, Piccoli P. Gibberellin production by bacteria and its involvement in plant growth promotion and yield increase. Applied Microbiology and Biotechnology. 2004;**65**(5):497-503

[53] Yamaguchi S. Gibberellin metabolism and its regulation.

Annual Review of Plant Biology. 2008;**59**:225-251

[54] Iqbal N, Nazar R, Khan MIR, Masood A, Khan NA. Role of gibberellins in regulation of source-sink relations under optimal and limiting environmental conditions. Current Science. 2011;**100**(7):998-1007

[55] Kumar B, Singh B. Effect of plant hormones on growth and yield of wheat irrigated with saline water. Annals of Agricultural Research. 1996;**17**:209-212

[56] Prakash L, Prathapasenan G. NaCl- and gibberellic acid-induced changes in the content of auxin and the activities of cellulase and pectin lyase during leaf growth in rice (*Oryza sativa*). Annals of Botany. 1990;**65**(3):251-257

[57] Moya J, Ros R, Picazo I. Heavy metal-hormone interactions in rice plants: Effects on growth, net photosynthesis, and carbohydrate distribution. Journal of Plant Growth Regulation. 1995;**14**(2):61

[58] Saleem M, Arshad M, Hussain S, Bhatti AS. Perspective of plant growth promoting rhizobacteria (PGPR) containing ACC deaminase in stress agriculture. Journal of Industrial Microbiology & Biotechnology. 2007;**34**(10):635-648

[59] Glick BR, editor. Promotion of plant growth by soil bacteria that regulate plant ethylene levels. In: Proceedings 33rd PGRSA Annual Meeting. 2007

[60] Hays DB, Do JH, Mason RE, Morgan G, Finlayson SA. Heat stress induced ethylene production in developing wheat grains induces kernel abortion and increased maturation in a susceptible cultivar. Plant Science. 2007;**172**(6):1113-1123

[61] Fahad S, Hussain S, Bano A, Saud S, Hassan S, Shan D, et al. Potential role of phytohormones and plant growth-promoting rhizobacteria in abiotic stresses: Consequences for changing environment. Environmental Science and Pollution Research. 2015;**22**(7):4907-4921

[62] Xing H, Tan L, An L, Zhao Z, Wang S, Zhang C. Evidence for the involvement of nitric oxide and reactive oxygen species in osmotic stress tolerance of wheat seedlings: Inverse correlation between leaf abscisic acid accumulation and leaf water loss. Plant Growth Regulation. 2004;**42**(1):61-68

[63] Flexas J, Bota J, Loreto F, Cornic G, Sharkey T. Diffusive and metabolic limitations to photosynthesis under drought and salinity in C3 plants. Plant Biology. 2004;**6**(03):269-279

[64] Yang J, Kloepper JW, Ryu C-M. Rhizosphere bacteria help plants tolerate abiotic stress. Trends in Plant Science. 2009;**14**(1):1-4

[65] Barnawal D, Bharti N, Pandey SS, Pandey A, Chanotiya CS, Kalra A. Plant growth-promoting rhizobacteria enhance wheat salt and drought stress tolerance by altering endogenous phytohormone levels and TaCTR1/TaDREB2 expression. Physiologia Plantarum. 2017;**161**(4):502-514

[66] Shahzad R, Khan AL, Bilal S, Waqas M, Kang S-M, Lee I-J. Inoculation of abscisic acid-producing endophytic bacteria enhances salinity stress tolerance in *Oryza sativa*. Environmental and Experimental Botany. 2017;**136**:68-77

[67] Miller MB, Bassler BL. Quorum sensing in bacteria. Annual Review of Microbiology. 2001;**55**(1):165-199

[68] Cheikh N, Jones RJ. Disruption of maize kernel growth and development by heat stress (role of cytokinin/abscisic acid balance). Plant Physiology. 1994;**106**(1):45-51

[69] Krishna P. Plant responses to heat stress. In: Plant Responses to Abiotic Stress. Berlin: Springer; 2003. pp. 73-101

[70] Heino P, Palva ET. Signal transduction in plant cold acclimation. In: Plant Responses to Abiotic Stress. Berlin: Springer; 2003. pp. 151-186

[71] Dolferus R, Ji X, Richards RA. Abiotic stress and control of grain number in cereals. Plant Science. 2011;**181**(4):331-341

[72] Egamberdiyeva D, Höflich G. Influence of growth-promoting bacteria on the growth of wheat in different soils and temperatures. Soil Biology and Biochemistry. 2003;**35**(7):973-978

[73] Ali SZ, Sandhya V, Grover M, Kishore N, Rao LV, Venkateswarlu B. Pseudomonas sp strain AKM-P6 enhances tolerance of sorghum seedlings to elevated temperatures. Biology and Fertility of Soils. 2009;**46**(1):45-55

[74] Mishra PK, Bisht SC, Ruwari P, Selvakumar G, Joshi GK, Bisht JK, et al. Alleviation of cold stress in inoculated wheat (*Triticum aestivum* L.) seedlings with psychrotolerant pseudomonads from NW Himalayas. Archives of Microbiology. 2011;**193**(7):497-513

[75] Blaylock AD. Soil Salinity, Salt Tolerance, and Growth Potential of Horticultural and Landscape Plants. University of Wyoming, Cooperative Extension Service, Department of Plant, Soil, and Insect Sciences; 1994

[76] Lambers H. Introduction, dryland salinity: A key environmental issue in southern Australia. Plant and Soil. 2003;**257**(2):5-7

[77] Munns R, James RA, Läuchli A. Approaches to increasing the salt tolerance of wheat and other cereals. Journal of Experimental Botany. 2006;**57**(5):1025-1043

[78] Ashraf M, Hasnain S, Berge O, Mahmood T. Inoculating wheat seedlings with exopolysaccharide-producing bacteria restricts sodium uptake and stimulates plant growth under salt stress. Biology and Fertility of Soils. 2004;**40**(3):157-162

[79] Hamdia MAE-S, Shaddad M, Doaa MM. Mechanisms of salt tolerance and interactive effects of *Azospirillum brasilense* inoculation on maize cultivars grown under salt stress conditions. Plant Growth Regulation. 2004;**44**(2):165-174

[80] Nadeem SM, Zahir ZA, Naveed M, Arshad M. Rhizobacteria containing ACC-deaminase confer salt tolerance in maize grown on salt-affected fields. Canadian Journal of Microbiology. 2009;**55**(11):1302-1309

[81] Chang P, Gerhardt KE, Huang X-D, Yu X-M, Glick BR, Gerwing PD, et al. Plant growth-promoting bacteria facilitate the growth of barley and oats in salt-impacted soil: Implications for phytoremediation of saline soils. International Journal of Phytoremediation. 2014;**16**(11):1133-1147

[82] Egamberdieva D. Alleviation of salt stress by plant growth regulators and IAA producing bacteria in wheat. Acta Physiologiae Plantarum. 2009;**31**(4):861-864

[83] Lambers H, Chapin III FS, Pons TL. Plant Physiological Ecology. New York: Springer Science & Business Media; 2008

[84] Barnabas B, Jager K, Feher A. The effect of drought and heat stress on reproductive processes in cereals. Plant, Cell & Environment. 2008;**31**(1):11-38

[85] Farooq M, Hussain M, Wahid A, Siddique K. Drought stress in plants:

An overview. In: Plant Responses to Drought Stress. Berlin: Springer; 2012. pp. 1-33

[86] Schussler J, Westgate M. Assimilate flux determines kernel set at low water potential in maize. Crop Science. 1995;**35**(4):1074-1080

[87] Yadav RS, Hash C, Bidinger F, Devos K, Howarth CJ. Genomic regions associated with grain yield and aspects of post-flowering drought tolerance in pearl millet across stress environments and tester background. Euphytica. 2004;**136**(3):265-277

[88] Ashraf M. Inducing drought tolerance in plants: Recent advances. Biotechnology Advances. 2010;**28**(1):169-183

[89] Hayat S, Hayat Q, Alyemeni MN, Wani AS, Pichtel J, Ahmad A. Role of proline under changing environments: A review. Plant Signaling & Behavior. 2012;**7**(11):1456-1466

[90] Casanovas EM, Barassi CA, Sueldo RJ. *Azospirillum* inoculation mitigates water stress effects in maize seedlings. Cereal Research Communications. 2002;**30**:343-350

[91] Vurukonda SSKP, Vardharajula S, Shrivastava M, SkZ A. Enhancement of drought stress tolerance in crops by plant growth promoting rhizobacteria. Microbiological Research. 2016;**184**:13-24

[92] Naveed M, Mitter B, Reichenauer TG, Wieczorek K, Sessitsch A. Increased drought stress resilience of maize through endophytic colonization by Burkholderia phytofirmans PsJN and *Enterobacter* sp. FD17. Environmental and Experimental Botany. 2014;**97**:30-39

[93] Sarig S, Okon Y, Blum A. Effect of *Azospirillum brasilense* inoculation on growth dynamics and hydraulic conductivity of Sorghum bicolor roots. Journal of Plant Nutrition. 1992;**15**(6-7):805-819

[94] Creus CM, Sueldo RJ, Barassi CA. Water relations and yield in *Azospirillum*-inoculated wheat exposed to drought in the field. Canadian Journal of Botany. 2004;**82**(2):273-281

[95] El-Komy H, Hamdia M, El-Baki GA. Nitrate reductase in wheat plants grown under water stress and inoculated with *Azospirillum* spp. Biologia Plantarum. 2003;**46**(2):281-287

[96] Wang W, Vinocur B, Altman A. Plant responses to drought, salinity and extreme temperatures: Towards genetic engineering for stress tolerance. Planta. 2003;**218**(1):1-14

[97] Shaharoona B, Naveed M, Arshad M, Zahir ZA. Fertilizer-dependent efficiency of pseudomonads for improving growth, yield, and nutrient use efficiency of wheat (*Triticum aestivum* L.). Applied Microbiology and Biotechnology. 2008;**79**(1):147-155

[98] Long SR. Rhizobium-legume nodulation: Life together in the underground. Cell. 1989;**56**(2):203-214

[99] Gutiérrez-Zamora M, Martınez-Romero E. Natural endophytic association between rhizobium etli and maize (*Zea mays* L.). Journal of Biotechnology. 2001;**91**(2-3):117-126

[100] Dobbelaere S, Croonenborghs A, Thys A, Ptacek D, Okon Y, Vanderleyden J. Effect of inoculation with wild type *Azospirillum brasilense* and *A. irakense* strains on development and nitrogen uptake of spring wheat and grain maize. Biology and Fertility of Soils. 2002;**36**(4):284-297

[101] Rosenblueth M, Ormeño-Orrillo E, López-López A, Rogel MA, Reyes-Hernández BJ, Martínez-Romero JC, et al.

Nitrogen fixation in cereals. Frontiers in Microbiology. 2018;**9**:1794-1800

[102] Van Dommelen A, Croonenborghs A, Spaepen S, Vanderleyden J. Wheat growth promotion through inoculation with an ammonium-excreting mutant of *Azospirillum brasilense*. Biology and Fertility of Soils. 2009;**45**(5):549-553

[103] Fox AR, Soto G, Valverde C, Russo D, Lagares A Jr, Zorreguieta Á, et al. Major cereal crops benefit from biological nitrogen fixation when inoculated with the nitrogen-fixing bacterium pseudomonas protegens Pf-5 X940. Environmental Microbiology. 2016;**18**(10):3522-3534

[104] Beatty PH, Good AG. Future prospects for cereals that fix nitrogen. Science. 2011;**333**(6041):416-417

[105] Rodríguez H, Fraga R. Phosphate solubilizing bacteria and their role in plant growth promotion. Biotechnology Advances. 1999;**17**(4-5):319-339

[106] Chabot R, Antoun H, Cescas MP. Growth promotion of maize and lettuce by phosphate-solubilizing rhizobium leguminosarum biovar. phaseoli. Plant and Soil. 1996;**184**(2):311-321

[107] Murty M, Ladha J. Influence of *Azospirillum* inoculation on the mineral uptake and growth of rice under hydroponic conditions. Plant and Soil. 1988;**108**(2):281-285

[108] Singh S, Kapoor K. Inoculation with phosphate-solubilizing microorganisms and a vesicular-arbuscular mycorrhizal fungus improves dry matter yield and nutrient uptake by wheat grown in a sandy soil. Biology and Fertility of Soils. 1999;**28**(2):139-144

[109] Castignetti D, Smarrelli J. Siderophores, the iron nutrition of plants, and nitrate reductase. FEBS Letters. 1986;**209**(2):147-151

[110] O'sullivan DJ, O'Gara F. Traits of fluorescent *Pseudomonas* spp. involved in suppression of plant root pathogens. Microbiology and Molecular Biology Reviews. 1992;**56**(4):662-676

[111] Rees DC, Howard JB. Nitrogenase: Standing at the crossroads. Current Opinion in Chemical Biology. 2000;**4**(5):559-566

[112] Rosconi F, Davyt D, Martínez V, Martínez M, Abin-Carriquiry JA, Zane H, et al. Identification and structural characterization of serobactins, a suite of lipopeptide siderophores produced by the grass endophyte *Herbaspirillum seropedicae*. Environmental Microbiology. 2013;**15**(3):916-927

[113] Egamberdiyeva D. The effect of plant growth promoting bacteria on growth and nutrient uptake of maize in two different soils. Applied Soil Ecology. 2007;**36**(2-3):184-189

[114] Coombs JT, Franco CM. Isolation and identification of actinobacteria from surface-sterilized wheat roots. Applied and Environmental Microbiology. 2003;**69**(9):5603-5608

[115] Yuwono T, Handayani D, Soedarsono J. The role of osmotolerant rhizobacteria in rice growth under different drought conditions. Australian Journal of Agricultural Research. 2005;**56**(7):715-721

[116] Solano BR, de la Iglesia MP, Probanza A, García JL, Megías M, Mañero FG, editors. Screening for PGPR to improve growth of *Cistus ladanifer* seedlings for reforestation of degraded mediterranean ecosystems. In: First International Meeting on Microbial Phosphate Solubilization. Springer; 2007

[117] Kim Y-C, Glick BR, Bashan Y, Ryu C-M. Enhancement of plant drought tolerance by microbes. In: Plant

Responses to Drought Stress. Berlin: Springer; 2012. pp. 383-413

[118] Suzuki N, Rivero RM, Shulaev V, Blumwald E, Mittler R. Abiotic and biotic stress combinations. The New Phytologist. 2014;**203**(1):32-43

[119] Lugtenberg BJ, Dekkers L, Bloemberg GV. Molecular determinants of rhizosphere colonization by pseudomonas. Annual Review of Phytopathology. 2001;**39**(1):461-490

[120] O'Callaghan M. Microbial inoculation of seed for improved crop performance: Issues and opportunities. Applied Microbiology and Biotechnology. 2016;**100**(13):5729-5746

[121] Nelson EB. Microbial dynamics and interactions in the spermosphere. Annual Review of Phytopathology. 2004;**42**:271-309

[122] Truyens S, Weyens N, Cuypers A, Vangronsveld J. Bacterial seed endophytes: Genera, vertical transmission and interaction with plants. Environmental Microbiology Reports. 2015;**7**(1):40-50

Chapter 5

The Infection Unit: An Overlooked Conceptual Unit for Arbuscular Mycorrhizal Function

Yoshihiro Kobae

Abstract

Most land plant species have their roots colonized by arbuscular mycorrhizal fungi (AMF). These symbiotic associations are often found in the roots of field crops. The biological basis and practical significance of this symbiosis have been extensively studied, and the molecular mechanisms underlying the initial colonization process and the nutrient exchange between the host plant and the AMF have been elucidated. However, developmental processes and turnover of elements of the mycorrhiza, and the resulting changes in mycorrhizal function, are not well understood. The enigmatic nature of the development-function relationship is probably due to the short life span of the infection unit, which has largely been overlooked in studies investigating mycorrhizal function at the macroscopic level. This paper outlines the concept of the infection unit and functional expression patterns in terms of the transient aspects of the micro-symbiont during its life cycle in this symbiosis.

Keywords: arbuscular mycorrhiza, functional molecular markers, infection unit, life cycle, live imaging

1. Introduction

The roots of approximately 95% of vascular land plant species, with the exception of some families (e.g., the Brassicaceae, Amaranthaceae, and Polygonaceae), are colonized by symbiotic fungi which form a mutualistic relationship (mycorrhiza) with the host plant roots [1, 2]. Approximately 10% of vascular plant species, mostly woody species, are colonized by ectomycorrhizal fungi, which belong to the *Basidiomycota*, *Ascomycota*, and (less commonly) *Zygomycota*, and the fungal hyphae grow extracellularly, forming a mantle of mycelium around the roots [3, 4]. Most of the remaining mycorrhizal fungi, with the exception of family-specific mycorrhiza, such as the ericoid or orchid-specific mycorrhizal fungi, colonize nonwoody plant species and belong to the subphylum *Glomeromycotina* [3, 5]. This fungal group is generally known as the arbuscular mycorrhizal fungi (AMF), because these fungi form highly branched hyphal structure, known as arbuscules, in root cortical cells, and spread intercellularly (*Arum*-type) or intracellularly (*Paris*-type) [6]. The formation of arbuscules has been regarded as the unique morphological feature of this symbiosis responsible for the nutrient (particularly phosphorus) exchange between the host plant and the AMF [7]. Arbuscule formation occurs in parallel with the expression of a specific cellular system to allow the accommodation of AMF within the root tissue and to achieve nutrient (e.g., phosphorus and nitrogen) uptake via AMF mycelia [8, 9]. Genetic disruption of genes in the symbiotic system of model plants (e.g., *Medicago truncatula*, *Lotus japonicus*, *Oryza sativa*) has revealed the nutritionally beneficial relationships between plants and AMF [10–12].

AMF can also colonize thalli of the early nonvascular land plants, namely, the liverworts and hornworts [13, 14]. Phylogenetic analyses indicated that symbiotic genes are present in the genomes of these early land plants, with the functions of the encoded proteins being conserved, suggesting that this symbiosis is phylogenetically widespread in plants [15]. AM symbiosis is beneficial for plants in relation not only to nutrition but also to the mitigation of biotic and abiotic stresses (e.g., resistance to pathogens, tolerance of drought and toxic element stress, increased biomass production, and secondary metabolite accumulation) [16–19]. Hence, the functionality of AM symbiosis can influence the productivity and quality of crops.

The effects on the host plant of AM symbiosis are commonly investigated by inoculation of the roots of the host plant with a (usually) single-species AMF in pot culture. However, field-grown inoculated roots can harbor AMF species other than the test AMF species, the functionality of which would not have been tested using an inoculation test because not all AMF in the roots can sporulate [20]. In addition, not all AMF which colonize roots are active and functional [21, 22], and the colonization process and the stability of the colonization by diverse AMF species in the same field-grown roots are unclear [23]. Overall, under field conditions, the functionality of AM symbiosis is probably based on unknown but highly dynamic associations between plants and a diverse range of AMF species. To better understand this complex association, in the current article, the colonization dynamics of AMF species in roots will be discussed. To understand the functional unit of the plant-AMF symbiosis, it is important to outline the concept of the colonization unit.

2. Latent colonization dynamics in arbuscular mycorrhizas

In 1905, an illustration by Gallaud showed that root cortical cells often contain "clumps" of arbuscules (**Figure 1**) [24]. Subsequent morphological examination of mycorrhizal roots at the cellular level suggested that this intracellular colonization may be ephemeral [25–27]. Following this, morphometric studies, coupled

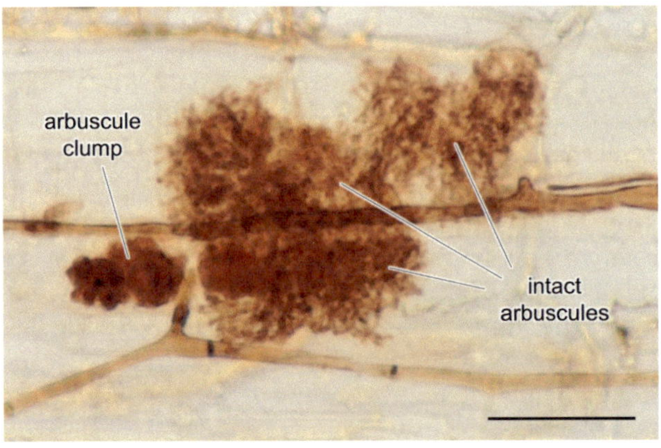

Figure 1.
Arbuscule has a short life span. Image of sunflower (Helianthus annuus) mycorrhizal roots grown in field soil. 3,3'-diaminobenzidine (DAB) staining with horseradish peroxidase (HRP)—Wheat germ agglutinin (WGA) [65]. Bar = 50 µm.

with electron microscopy, calculated that the life span of an intact arbuscule in several plant species was a minimum of 2.5 days [28, 29]. Such a limited life span of the units of intracellular colonization by AMF can be generalized because these arbuscule clumps have been observed in many plant species, including the relatively primitive nonvascular plant, the liverwort [13]. Why does this mutually beneficial association exhibit such short-lived units of intracellular colonization? Unfortunately, our understanding of this phenomenon has not increased much since the first observation of arbuscule clumps by Gallaud more than 110 years ago.

Advances in the forward and reverse genetic approaches available to decipher the molecular mechanism for the AMF colonization process have shown that several signaling mutants of model plants exhibited compromised AMF epidermal penetration and altered chemical and cellular crosstalk in the initial stages between plants and AMF [10, 30]. Recently, several mutant lines exhibited suppressed intraradical colonization, forming prematurely senescent or stunted arbuscules [11, 31]. Some plant genes have been implicated in the cellular process of arbuscule degeneration [32–35]. The degeneration process of arbuscules has also been shown to be related to colonization level, illustrating the importance of arbuscule life cycle to the development of mycorrhiza. In addition, live imaging of the green fluorescent protein (GFP)—symbiotic phosphate transporter (PT11) fusion protein in the roots of mycorrhized rice seedlings—also revealed the limited life span of the units of intracellular colonization, in which the rapid collapse of arbuscules was observed [36]. The mechanism of mycorrhization, coupled with such a short life span of the individual colonization unit, has been addressed by continuous (long-term) live imaging of the symbiotic marker secretory carrier membrane protein (SCAMP) [37], where successive de novo colonizations underlie AM development (SubSection 2.2). These findings emphasize the importance of the life cycle of colonization when we consider the dynamics of AM functionality.

2.1 Concept of the infection unit

In soils, there are many different AMF species, and no strict host-AMF specificity has been observed [38, 39]. Accordingly, under field conditions, co-colonization of the same root by multiple AMF species can occur [2]. It is likely that the functionality of these diverse AMFs is not the same, and colonization by each AMF may last only a short time. Therefore, to correctly characterize the functionality of field mycorrhizas, the colonization process, the dynamics, and the functionality of the diverse AMFs in the roots need to be understood.

When the fungal spore germ tubes approach the root surface, chemical crosstalk occurs between the roots and the AMF hyphae, triggering the molecular and cellular remodeling process necessary for hyphal entry into the roots [11]. Coinciding with this pre-symbiotic crosstalk, AMF hyphae around the roots are often highly branched, giving rise to a characteristic cascade-like mycelium, composed of lateral branches [2]. Several plant mutants exhibiting disruption of the early signaling process fail to allow hyphal penetration of the epidermal layer [10, 11], suggesting the importance of this initial mutual recognition process.

After hyphopodia are formed, hyphae penetrate the rhizodermal layer and grow longitudinally in the root cortex. Short branches from the longitudinally extending hyphae penetrate the cortical cell walls and branch dichotomously in the cell lumen to give rise to arbuscules. Importantly, the maximum elongation in the cortex of hyphal structures derived from a single or a few hyphal penetrations of the epidermis are reported to be up to 20 mm [2]. In rice seedlings, however, the maximum elongation in the cortex of hyphal structures derived from the entry point is only 0.5 mm [37]. In general, the area occupied by each colonization unit derived from a single epidermal entry is difficult to recognize, because intraradical hyphae derived from different entry points immediately overlap one another to form a larger colonization area within the cortex. The rates of growth of intraradical colonization are reported to range from 0.13 to 1.22 mm/day [2]. In the live imaging of mycorrhizal rice roots, the rate varied from 0.42 to 1.68 mm/day [36]. Although the maximum length of intraradical colonization derived from one or a few penetrations varied greatly among studies, these independent colonies of mycelia are called the "infection unit" (**Figure 2**) [25].

2.2 Development and functionality of mycorrhizas: an infection unit-based view

The development of the mycorrhiza and the dynamics of the infection unit are tightly linked processes during mycorrhization. It is also likely that the functionality of mycorrhiza in the field is also variable, depending on the functionality of the infection unit that is derived from different AMF species. Live in situ imaging revealed that multiple infection units overlap to form a larger infection ("colonized region")

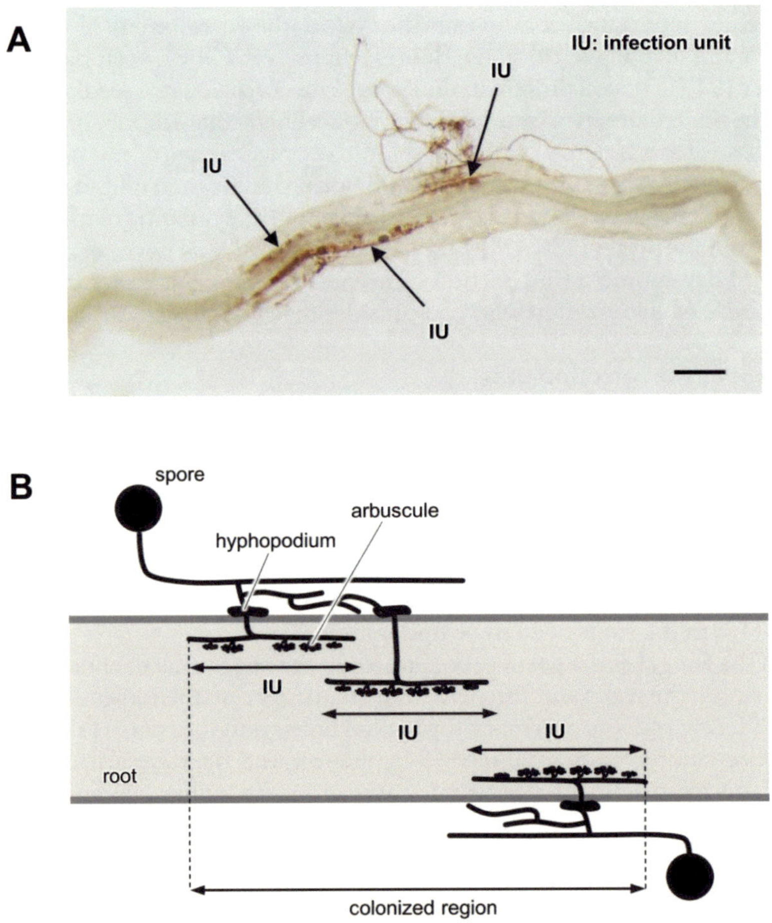

Figure 2.
The relationship between infection unit and mycorrhizal development. (A) Image of Lotus japonicus seedlings grown in field soil taken at 14 days post plantation. 3,3'-diaminobenzidine (DAB) staining with horseradish peroxidase (HRP)—Wheat germ agglutinin (WGA) [65]. Bar = 200 μm. (B) Model diagram of mycorrhization process. AMF spores germinate in the soil, and germ tubes approach the root surface and form a hyphopodium. The young infection unit, comprising an internal mycelium arising from one hyphal entry, grows and develops new arbuscules at the infection fronts. In many cases, new infection units develop immediately adjacent to established infection units. Colonized regions that were bound by two infection fronts, comprising intercellular and intracellular hyphae, develop through the successive formation of infection units. Arbuscules collapse from near the hyphopodia as a result of their short life span.

within a few days [36, 37, 40], making the delimitation of each infection unit experimentally difficult. Coupled with the difficulty of differentiating between different AMF species in the roots, using morphological or molecular approaches [41, 42], the dynamic process of mycorrhization has probably hampered progress in characterizing the infection unit-based functionality of field mycorrhizas. Live imaging of rice mycorrhizal roots revealed that the expansion of the colonized region occurs in concert with successive de novo formation of multiple infection units (**Figure 2**) [37].

However, a better understanding of the dynamics of the mycorrhization process depends substantially on the quality and the ease of imaging of the hyphae within the roots. Traditionally, cytological studies of the colonization process have been performed with chemical staining of fixed (i.e., dead) AMF structures or in situ visualization of their enzymatic activities. In recent studies, however, imaging of non-fixed root samples by means of fluorescent molecular markers of the symbiotic process has been used to improve our understanding of colonization dynamics at the cellular level. As many parts of the molecular mechanism implicated in AM symbiosis are common to nodule symbiosis, several molecular markers are available in the model legume plants, *M. truncatula* and *L. japonicus* [30]. However, live imaging of mycorrhizas in model legumes by means of fluorescent markers is difficult due to the presence of highly autofluorescent materials in the root tissue and the presence of thick (multiple) cortical cell layers that decrease transparency [43]. Such poorly transparent root tissues are not particularly useful for macroscopic imaging of the dynamics of the infection unit.

Rice (*O. sativa*) is commonly grown in paddy fields, where it is rarely colonized with AMF. However, rice roots are colonized with AMF under semidry conditions. The root systems of leguminous species have cambium, and the diameters of primary and lateral roots are generally uniformly large. On the other hand, grass roots do not have cambium, and lateral roots are smaller in diameter than the primary (crown) root. In rice seedlings, lateral roots with a few cortical layers are the main site of AMF colonization, and the average diameter of roots is less than 200 μm [36, 44]. In addition, the concentrations of autofluorescent root materials are quite low, and some symbiotic molecular markers are available for live imaging [36, 37, 45].

Transcriptome analysis of mycorrhized rice roots revealed that an AM-specific marker gene of rice, *AM42*, which encodes a SCAMP, is specifically expressed in mycorrhized roots [46, 47]. A GFP-tagged SCAMP protein was localized in the endomembrane systems of colonized cells and even in cells with collapsed arbuscules, allowing live imaging, coupled with GFP-SCAMP, to evaluate the colonization and recolonization sequences. Live imaging of GFP-SCAMP revealed that the average lifetime of intact arbuscules was 1–2 days. Cortical cells with collapsed arbuscules were rarely recolonized, whereas new colonizations occurred in close proximity to cells containing collapsed arbuscules, contributing to the expansion of the colonized region. Collapsed arbuscule-containing cells are intact [2]; however, colonization spread readily into an uncolonized region of roots but sparsely into a previously colonized region, suggesting that successive formation of new infection units is required for continuous mycorrhization [37, 48]. It is unlikely that the collapse and the presumed digestion of arbuscules play a significant role in nutrient transfer from fungus to plant [2].

The concept that mycorrhization is linked to the successive formation of infection units is supported by the observation that decreased hyphopodium formation leads to decreased mycorrhization. Under low-phosphate conditions, roots secrete strigolactones (SLs), which are carotenoid-derived phytohormones. A chemical analog of SLs, GR-24, activated mitochondrial respiratory activity and facilitated hyphal branching of *Gigaspora margarita* or *G. rosea* under in vitro conditions [49–51]. In SL biosynthesis-defective rice mutants, the hyphal branching of a model AMF, *Rhizophagus irregularis*, around the roots (rhizospheric hyphal branching) was

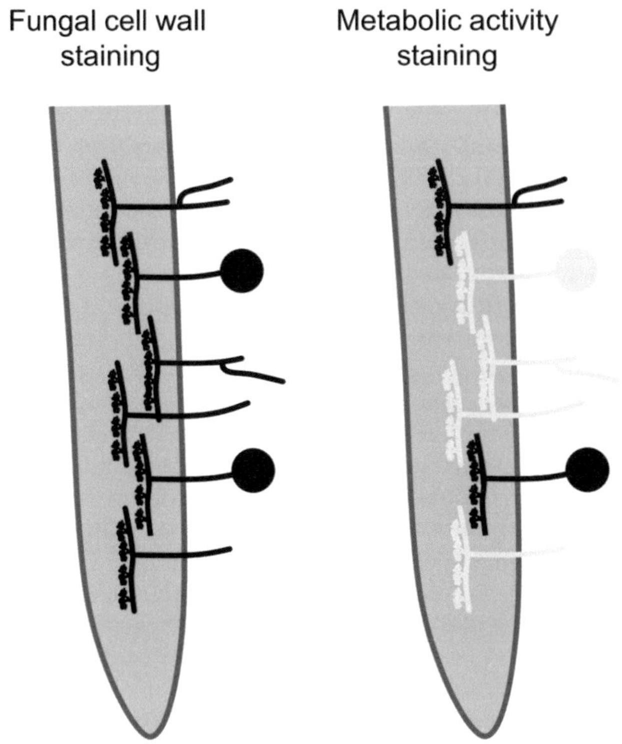

Figure 3.
Not all infection units revealed with fungal cell wall staining are metabolically active. Roots colonized with native AM fungi in field soils were subjected to cell wall (chitin) staining and vital staining to detect the presence of AMF and metabolically active AMF, respectively [60]. Vital staining, which histochemically visualizes the activity of succinate dehydrogenase (SDH), a tricarboxylic acid cycle enzyme in AM fungi, using the reduction of nitroblue tetrazolium (NBT) into insoluble formazan, detects metabolically active colonization [66]. The number of infection units detected by vital staining was lower than that determined by cell wall staining. In this analysis, rice (Oryza sativa L.) was used as the host plant because (i) the morphology of the development of infection units is well understood [36, 37]; (ii) active infection units rarely coalesce in roots [37], probably due to the small number of cortical cell layers [67, 68]; and (iii) the vital staining is convenient for detecting a single infection unit [36, 44].

normal [48]; however, in the SL biosynthesis-defective mutants of pea, tomato, and rice, the percentage root length colonization was significantly reduced [52–55]. In the rice SL biosynthesis mutants, the formation of the hyphopodium was delayed, compared with the wild type, but intraradical colonization was normal, indicating that the early formation of infection units, initiated by timely hyphal entry into epidermal cells, is necessary for the normal development of a colonized region [48].

In the field, multiple AMFs, from different species, can co-colonize roots, with multiple AMF species being detectable in only a 1-cm-long root fragment [56]. Thus, field roots can be regarded as a mosaic of the various functionalities of the different AMFs [57]; alternatively, only a portion of the AMF infection unit colonizing the roots may temporarily contribute to particular functions in response to specific environmental conditions [58]. However, the dynamics of the functionality of the respective AMFs in field roots have been little studied. Abiotic and biotic factors may influence the AMF composition, at least over a long time period [59], but the short-term effects of such environmental factors may also influence the various active (functional) AMFs as well as the functionality of the mycorrhiza as an entirety under field conditions. For example, not all infection units containing fine-branched arbuscules in roots grown in field soils are metabolically active (**Figure 3**) [60]. Further study will be needed to understand the functional dynamics of field mycorrhizas by considering infection unit-based colonization dynamics.

3. Conclusion

Our understanding of the molecular mechanism underlying the plant nutrient uptake system has been greatly advanced by the use of molecular studies and genetics. These studies have been based almost entirely on the plant alone. However, the plant nutrient uptake system is largely dependent on the largely undiscovered functionalities of diverse soil microorganisms. Roots of plants in the field are generally colonized with a range of different AMFs, but the functionality of these individual species is largely unknown. In addition, the genetic structure of AMF is quite enigmatic [41]. Recent studies into single-nucleus sequencing of some AMF culture lines demonstrated the presence of not only homokaryons but also dikaryons and heterokaryons [61, 62]. Furthermore, long-read whole-genome sequencing of *R. irregularis* DAOM197198 indicated that the genome contained 10 different rDNA sequences that were scattered (i.e., non-tandem repeats) around the chromosome [63]. These findings mitigate against the use of rDNA sequences to identify individual AM species in field AMF infection units, as a one-to-one relationship may not be applicable to the rDNA sequences and the genetic identities of the individual component species.

As mentioned before, AM infection units have a short life span and collapse within a few days, at least in the live imaging of rice seedlings [36, 37]. The development of mycorrhizal roots is associated with the turnover and de novo colonization by new AMFs, providing the opportunity to allow different AMF species to colonize the roots, depending on the context (environmental factors, plant growth stage, nutrition, etc.). Accordingly, field mycorrhizal roots can comprise multiple functionalities with different AMF species.

Plant breeding programs are not able to select for the genomic properties of plants adapted to all field conditions (soil type, water content, nutrient level, climate, etc.). On the other hand, plant roots closely interact with the native fungal partners that may have genetically recorded beneficial traits for adapting to the local environment. The specific phenotype (functionality) of native AMFs may be conferred by accessory genes that are not shared by all members of a species [64]. However, due to the complexity of the genetic basis of AMF individuals, it is difficult to understand which AMF genes are really functional in the roots. In a model plant (e.g., rice), the thin root cortex would allow us to isolate the genetic information of AMF individuals in the form of the infection unit. Furthermore, rice roots are technically suitable for detecting metabolic activities by means of vital staining. Furthermore, transgenic rice producing fluorescent molecular markers (e.g., phosphate transporter, GFP) is available to assess the functionality of the AMFs in situ. Thus, future studies should focus on the functionality of field AMF individuals, with emphasis on the genetic information and the dynamic functionality.

Acknowledgements

This work was supported partly by the Japan Science and Technology Agency [ACCEL grant No. JPMJAC1403].

Conflict of interest

The author declares that they have no conflict of interest.

Author details

Yoshihiro Kobae
Laboratory of Crop Nutrition, Department of Sustainable Agriculture, Rakuno Gakuen University, Ebetsu, Hokkaido, Japan

*Address all correspondence to: kobae@rakuno.ac.jp

References

[1] Trappe JM. Selection of fungi for ectomycorrhizal inoculation in nurseries. Annual Review of Phytopathology. 1977;**15**:203-222

[2] Smith SE, Read DJ. Mycorrhizal Symbiosis. Cambridge, UK: Academic Press; 2008

[3] Wang B, Qiu Y. Phylogenetic distribution and evolution of mycorrhizas in land plants. Mycorrhiza. 2006;**16**:299-363. DOI: 10.1007/s00572-005-0033-6

[4] Balestrini R, Bonfante P. Cell wall remodeling in mycorrhizal symbiosis: A way towards biotrophism. Frontiers in Plant Science. 2014;**5**:237

[5] Strullu-Derrien C, Selosse MA, Kenrick P, Martin FM. The origin and evolution of mycorrhizal symbioses: From palaeomycology to phylogenomics. The New Phytologist. 2018;**220**:1012-1030

[6] Dickson S, Smith FA, Smith SE. Structural differences in arbuscular mycorrhizal symbioses: More than 100 years after Gallaud, where next? Mycorrhiza. 2007;**17**:375-393

[7] Smith SE, Smith FA. Roles of arbuscular mycorrhizas in plant nutrition and growth: New paradigms from cellular to ecosystem scales. Annual Review of Plant Biology. 2011;**62**:227-250

[8] Bücking H, Liepold E, Ambilwade P. The role of the mycorrhizal symbiosis in nutrient uptake of plants and the regulatory mechanisms underlying these transport processes. Plant Science. 2012;**4**:108-132. DOI: 10.5772/52570

[9] Luginbuehl LH, Oldroyd GE. Understanding the arbuscule at the heart of endomycorrhizal symbioses in plants. Current Biology. 2017;**27**:R952-R963

[10] MacLean AM, Bravo A, Harrison MJ. Plant signaling and metabolic pathways enabling arbuscular mycorrhizal symbiosis. The Plant Cell. 2017;**29**:2319-2335

[11] Choi J, Summers W, Paszkowski U. Mechanisms underlying establishment of arbuscular mycorrhizal symbioses. Annual Review of Phytopathology. 2018;**56**:135-160

[12] Lanfranco L, Fiorilli V, Gutjahr C. Partner communication and role of nutrients in the arbuscular mycorrhizal symbiosis. The New Phytologist. 2018;**220**:1031-1046. DOI: 10.1111/nph.15230

[13] Ligrone R, Carafa A, Lumini E, Bianciotto V, Bonfante P, Duckett JG. Glomeromycotean associations in liverworts: A molecular, cellular, and taxonomic analysis. American Journal of Botany. 2007;**94**:1756-1777

[14] Rimington WR, Pressel S, Duckett JG, Field KJ, Read DJ, Bidartondo MI. Ancient plants with ancient fungi: Liverworts associate with early-diverging arbuscular mycorrhizal fungi. Proceedings of the Royal Society B. 2018;**285**:pii: 20181600. DOI: 10.1098/rspb.2018.1600

[15] Delaux PM, Radhakrishnan GV, Jayaraman D, Cheema J, Malbreil M, Volkening JD, et al. Algal ancestor of land plants was preadapted for symbiosis. Proceedings of the National Academy of Sciences of the United States of America. 2015;**112**:13390-13395

[16] Engel R, Szabó K, Abrankó L, Rendes K, Füzy A, Takács T. Effect of arbuscular mycorrhizal fungi on the growth and polyphenol profile of marjoram, lemon balm, and Marigold. Journal of Agricultural and Food Chemistry. 2016;**64**:3733-3742

[17] Rozpądek P, Rąpała-Kozik M, Wężowicz K, Grandin A, Karlsson S, Ważny R, et al. Arbuscular mycorrhiza improves yield and nutritional properties of onion (*Allium cepa*). Plant Physiology and Biochemistry. 2016;**107**:264-272

[18] Srivastava S, Conlan XA, Cahill DM, Adholeya A. *Rhizophagus irregularis* as an elicitor of rosmarinic acid and antioxidant production by transformed roots of *Ocimum basilicum* in an in vitro co-culture system. Mycorrhiza. 2016;**26**:919-930

[19] Plett JM, Martin FM. Know your enemy, embrace your friend: Using omics to understand how plants respond differently to pathogenic and mutualistic microorganisms. The Plant Journal. 2018;**93**:729-746

[20] Ohsowski BM, Zaitsoff PD, Öpik M, Hart MM. Where the wild things are: Looking for uncultured glomeromycota. The New Phytologist. 2014;**204**:171-179

[21] Vierheilig H, Ocampo JA. Relationship between SDH-activity and VA-mycorrhizal infection. Agriculture, Ecosystems and Environment. 1990;**29**:439-442

[22] Hamel C, Fyles H, Smith DL. Measurement of development of endomycorrhizal mycelium using three different vital stains. The New Phytologist. 1990;**115**:297-302

[23] Kobae Y. Dynamic phosphate uptake in arbuscular mycorrhizal roots under field conditions. Frontiers in Environmental Science. 2019;**6**:159

[24] Gallaud I. Études sur les mycorrhizes endotrophes. Revue Générale de Botanique. 1905;**17**:5-48, 66-83, 123-136, 223-239, 313-325, , 378, 425-433, 479-500

[25] Cox G, Sanders F. Ultrastructure of the host–fungus interface in a vesicular-arbuscular mycorrhiza. The New Phytologist. 1974;**73**:901-912

[26] Cox G, Tinker PB. Translocation and transfer of nutrients in vesicular-arbuscular mycorrhizas. The New Phytologist. 1976;**77**:371-378

[27] Bonfante-Fasolo P. Anatomy and morphology of VA mycorrhizae. In: Powell CL, Bagyaraj DJ, editors. VA Mycorrhizas. Boca Raton, FL: CRC Press; 1984. pp. 5-33

[28] Alexander T, Meier R, Toth R, Weber HC. Dynamics of arbuscule development and degeneration in mycorrhizas of *Triticum aestivum* L. and *Avena sativa* L. with reference to *Zea mays* L. The New Phytologist. 1988;**110**:363-370

[29] Alexander T, Toth R, Meier R, Weber HC. Dynamics of arbuscule development and degeneration in onion, bean, and tomato with reference to vesicular-arbuscular mycorrhizae in grasses. Canadian Journal of Botany. 1989;**67**:2505-2513

[30] Zipfel C, Oldroyd GE. Plant signalling in symbiosis and immunity. Nature. 2017;**543**:328

[31] Pimprikar P, Gutjahr C. Transcriptional regulation of arbuscular mycorrhiza development. Plant and Cell Physiology. 2018;**59**:678-695

[32] Javot H, Penmetsa RV, Terzaghi N, Cook DR, Harrison MJ. A *Medicago truncatula* phosphate transporter indispensable for the arbuscular mycorrhizal symbiosis. Proceedings of the National Academy of Sciences of the United States of America. 2007;**104**:1720-1725

[33] Yang SY, Grønlund M, Jakobsen I, Grotemeyer MS, Rentsch D, Miyao A, et al. Nonredundant regulation of rice arbuscular mycorrhizal symbiosis by two members of the *phosphate transporter1* gene family. The Plant Cell. 2012;**24**:4236-4251

[34] Floss DS, Gomez SK, Park HJ, MacLean AM, Müller LM, Bhattarai KK, et al. A transcriptional program for arbuscule degeneration during AM symbiosis is regulated by MYB1. Current Biology. 2017;**27**:1206-1212

[35] Li C, Zhou J, Wang X, Liao H. A purple acid phosphatase, *GmPAP33*, participates in arbuscule degeneration during arbuscular mycorrhizal symbiosis in soybean. Plant, Cell and Environment. 2019;**42**(6):2015-2027. DOI: 10.1111/pce.13530

[36] Kobae Y, Hata S. Dynamics of periarbuscular membranes visualized with a fluorescent phosphate transporter in arbuscular mycorrhizal roots of rice. Plant and Cell Physiology. 2010;**51**:341-353

[37] Kobae Y, Fujiwara T. Earliest colonization events of *Rhizophagus irregularis* in rice roots occur preferentially in previously uncolonized cells. Plant and Cell Physiology. 2014;**55**:1497-1510

[38] Mosse B. Specificity in VA mycorrhizas. In: Endomycorrhizas. London, UK: Academic Press; 1975. pp. 469-484

[39] McGonigle TP, Fitter AH. Ecological specificity of vesicular-arbuscular mycorrhizal associations. Mycological Research. 1990;**94**:120-122

[40] Buwalda JG, Stribley DP, Tinker PB. The development of endomycorrhizal root systems. V. The detailed pattern of development of infection and the control of infection level by host in young leek plants. The New Phytologist. 1984;**96**:411-427

[41] Bruns TD, Corradi N, Redecker D, Taylor JW, Öpik M. Glomeromycotina: What is a species and why should we care? New Phytologist. 2018;**220**: 963-967. DOI: 10.1111/nph.14913

[42] Öpik M, Davison J. Uniting species- and community-oriented approaches to understand arbuscular mycorrhizal fungal diversity. Fungal Ecology. 2016;**24**:106-113

[43] Kobae Y, Tamura Y, Takai S, Banba M, Hata S. Localized expression of arbuscular mycorrhiza-inducible ammonium transporters in soybean. Plant and Cell Physiology. 2010;**51**:1411-1415

[44] Kobae Y, Gutjahr C, Paszkowski U, Kojima T, Fujiwara T, Hata S. Lipid droplets of arbuscular mycorrhizal fungi emerge in concert with arbuscule collapse. Plant and Cell Physiology. 2014;**55**:1945-1953

[45] Roth R, Hillmer S, Funaya C, Chiapello M, Schumacher K, Lo Presti L, et al. Arbuscular cell invasion coincides with extracellular vesicles and membrane tubules. Nature Plants. 2019;**5**:204

[46] Güimil S, Chang HS, Zhu T, Sesma A, Osbourn A, Roux C, et al. Comparative transcriptomics of rice reveals an ancient pattern of response to microbial colonisation. Proceedings of the National Academy of Sciences of the United States of America. 2005;**102**:8066-8070

[47] Gutjahr C, Banba M, Croset V, An K, Miyao A, An G, et al. Arbuscular mycorrhiza-specific signaling in rice transcends the common symbiosis signaling pathway. The Plant Cell. 2008;**20**:2989-3005

[48] Kobae Y, Kameoka H, Sugimura Y, Saito K, Ohtomo R, Fujiwara T, et al. Strigolactone biosynthesis genes of rice are required for the punctual entry of arbuscular mycorrhizal fungi into the roots. Plant and Cell Physiology. 2018;**59**:544-553

[49] Akiyama K, Matsuzaki KI, Hayashi H. Plant sesquiterpenes induce hyphal

branching in arbuscular mycorrhizal fungi. Nature. 2005;**435**:824

[50] Besserer A, Puech-Pagès V, Kiefer P, Gomez-Roldan V, Jauneau A, Roy S, et al. Strigolactones stimulate arbuscular mycorrhizal fungi by activating mitochondria. PLoS Biology. 2006;**4**:e226

[51] Salvioli A, Ghignone S, Novero M, Navazio L, Venice F, Bagnaresi P, et al. Symbiosis with an endobacterium increases the fitness of a mycorrhizal fungus, raising its bioenergetic potential. The ISME Journal. 2016;**10**:130-144

[52] Gomez-Roldan V, Fermas S, Brewer PB, Puech-Pagès V, Dun EA, Pillot JP, et al. Strigolactone inhibition of shoot branching. Nature. 2008;**455**:189-194

[53] Koltai H, LekKala SP, Bhattacharya C, Mayzlish-Gati E, Resnick N, Wininger S, et al. A tomato strigolactone-impaired mutant displays aberrant shoot morphology and plant interactions. Journal of Experimental Botany. 2010;**61**:1739-1749

[54] Vogel JT, Walter MH, Giavalisco P, Lytovchenko A, Kohlen W, Charnikhova T, et al. SlCCD7 controls strigolactone biosynthesis, shoot branching and mycorrhiza-induced apocarotenoid formation in tomato. The Plant Journal. 2010;**61**:300-311

[55] Yoshida S, Kameoka H, Tempo M, Akiyama K, Umehara M, Yamaguchi S, et al. The D3 F-box protein is a key component in host strigolactone responses essential for arbuscular mycorrhizal symbiosis. The New Phytologist. 2012;**196**:1208-1216

[56] van Tuinen D, Jacquot E, Zhao B, Gollotte A, Gianinazzi-Pearson V. Characterization of root colonization profiles by a microcosm community of arbuscular mycorrhizal fungi using 25S rDNA-targeted nested PCR. Molecular Ecology. 1988;**7**:879-887

[57] Jansa J, Smith FA, Smith SE. Are there benefits of simultaneous root colonization by different arbuscular mycorrhizal fungi? The New Phytologist. 2008;**177**:779-789

[58] Compant S, van der Heijden MG, Sessitsch A. Climate change effects on beneficial plant-microorganism interactions. FEMS Microbiology Ecology. 2010;**73**:197-214

[59] Xu X, Chen C, Zhang Z, Sun Z, Chen Y, Jiang J, et al. The influence of environmental factors on communities of arbuscular mycorrhizal fungi associated with *Chenopodium ambrosioides* revealed by MiSeq sequencing investigation. Scientific Reports. 2017;**7**:45134

[60] Kobae Y, Ohtomo R, Oka N, Morimoto S. A simple model system for identifying arbuscular mycorrhizal fungal taxa that actively colonize rice (*Oryza sativa* L.) roots grown in field soil. Soil Science & Plant Nutrition. 2017;**63**:29-36

[61] Ropars J, Toro KS, Noel J, Pelin A, Charron P, Farinelli L, et al. Evidence for the sexual origin of heterokaryosis in arbuscular mycorrhizal fungi. Nature Microbiology. 2016;**1**:16033

[62] Chen EC, Mathieu S, Hoffrichter A, Sedzielewska-Toro K, Peart M, Pelin A, et al. Single nucleus sequencing reveals evidence of inter-nucleus recombination in arbuscular mycorrhizal fungi. eLife. 2018;**7**:e39813

[63] Maeda T, Kobayashi Y, Kameoka H, Okuma N, Takeda N, Yamaguchi K, et al. Evidence of non-tandemly repeated rDNAs and their intragenomic heterogeneity in *Rhizophagus irregularis*. Communications Biology. 2018;**1**:87

[64] Mathieu S, Cusant L, Roux C, Corradi N. Arbuscular mycorrhizal fungi: Intraspecific diversity and pangenomes. The New Phytologist. 2018;**220**:1129-1134

[65] Kobae Y, Ohtomo R. An improved method for bright-field imaging of arbuscular mycorrhizal fungi in plant roots. Soil Science & Plant Nutrition. 2016;**62**:27-30. DOI: 10.1080/00380768.2015.1106923

[66] MacDonald RM, Lewis M. The occurrence of some acid phosphatases and dehydrogenases in the vesicular-arbuscular mycorrhizal fungus *Glomus mosseae*. New Phytologist. 1978;**80**:135-141. DOI: 10.1111/nph.1978.80.issue-1

[67] Fiorilli V, Vallino M, Biselli C, Faccio A, Bagnaresi P, Bonfante P. Host and non-host roots in rice: Cellular and molecular approaches reveal differential responses to arbuscular mycorrhizal fungi. Frontiers in Plant Science. 2015;**6**:636. DOI: 10.3389/fpls.2015.00636

[68] Gutjahr C, Sawers RJH, Marti G, Andrés-Hernández L, Yang SY, Casieri L, et al. Transcriptome diversity among rice root-types during asymbiosis and interaction with arbuscular mycorrhizal fungi. Proceedings of the National Academy of Sciences of the United States of America. 2015;**112**:6754-6759. DOI: 10.1073/pnas.1504142112

Section 3

Metabolites and Human Health

Chapter 6

Salted Radish Root Biology during Food Processing

Hiroki Matsuoka, Kei Kumakura, Taito Kobayashi,
Wataru Kobayashi and Asaka Takahashi

Abstract

White radish root (daikon) is an important vegetable in Japanese food culture and has spawned the development of various cooking and processing methods. takuan-zuke is the major processed food derived from daikon. Takuan-zuke is prepared by the dehydration of the root using a salt-press or a weighted stone, or by sun-drying, followed by salt-aging using salt or salted rice bran. The color of takuan-zuke changes to yellow during salt-aging. We determined the effects of dehydration and salt-aging on the metabolism of daikon using takuan-zuke. In the yellowing reaction, the generation of daikon isothiocyanate was significant, requiring a temperature of ≥10°C and pH ≥5. The color change of the sun-dried takuan-zuke was the most significant. Moreover, we investigated the nutritional characteristics of takuan-zuke. In the sun-dried daikon, metabolism progressed for 3 weeks during drying, with increase in the concentrations of γ-aminobutyric acid (GABA) and proline as well as drying stress metabolites. In the salt-pressed daikon, GABA concentrations temporarily increased due to osmotic stress but then decreased on metabolic inhibition by salt permeation. In addition, no change in the concentration of proline was observed under salt-press conditions. The results showed a marked difference between the stress response of the living and processed root.

Keywords: salted radish root, takuan-zuke, *Tsukemono*, isothiocyanates, metabolome analysis, GABA

1. Introduction

Daikon (Japanese white radish, *Raphanus sativus* L. var. longipinnatus Bailey) is an important vegetable in Japanese food culture (*Washoku*). The literal translation of daikon to English is big (or giant) root. In Japan, the weight of one root is 1.5–2 kg. Daikon is an indispensable ingredient of *Washoku* and has been used in salads, simmered dishes (*Nimono*), pickles (*Tsukemono*), and in the form of grated radish root (*Daikon oroshi*). In Japan, pickled daikon (salted radish root), which has been salt-aged for several months with salt or salted rice bran, is called takuan-zuke. It is presumed that the original form of takuan-zuke was made over 300 years ago. Currently, takuan-zuke is made by dehydrating raw daikon by sun-drying or salt-pressing followed by aging with salt or salty bran. The amount of salt and the length of salt-aging are dependent on the shipping time. With the modernization

Isothiocyanate	Relative Amount
CH_3–$(CH_2)_4$–N=C=S	+
$(CH_3)_2$–CH–$(CH_2)_2$–N=C=S	++
CH_3–$(CH_2)_5$–N=C=S	+
CH_3S–$(CH_2)_3$–N=C=S	++
C_6H_5–CH_2–N=C=S	+
CH_3S–CH=CH–$(CH_2)_2$–N=C=S	+++++
CH_3S–$(CH_2)_4$–N=C=S	++
CH_3S–$(CH_2)_5$–N=C=S	+++

Table 1.
Volatile isothiocyanates in Japanese radish roots [1]. Relative data was obtained by gas chromatograph with a flame ionization detector [1]. Symbols: +, <1%; ++, <1–5%; +++, 5–10%; +++++, >70%.

of pickles, the term "Genboku" refers to aged takuan-zuke, while low-salted radish roots stored at low temperatures in seasoning liquid are currently known as takuan-zuke. In Japan, takuan-zuke is a representative pickle, eaten as a side dish to steamed rice (*Gohan*), while in Europe and the United States, pickles are used as ingredients for cooking.

The flavor of raw daikon and its processed products is derived from a characteristic, pungent chemical component [1]. As shown in **Table 1**, this component, 4-methylthio-3-butenyl isothiocyanate (MTBITC), is formed by the enzymatic conversion of glucoraphasatin (GRH), which is a glucosinolate (GSL), by myrosinase (**Figure 1**). Isothiocyanates (ITCs) act as protective agents against pests in plants. In processed radish roots, such as takuan-zuke, MTBITC contributes by imparting its characteristic flavor and color [2–4]. Nakamura et al. reported the presence of 37–420 μmol/100 g of MTBITC and 280–1270 μmol/100 g of GRH in raw radishes, based on analysis results from a total of 83 samples from 7 varieties [5].

Owing to their antimicrobial properties, naturally occurring GSLs and ITCs have been studied for a long time, with 132 types identified by 2011 [6–9]. Recently, the role of ITC in human redox regulation and the activation of detoxification enzymes which act against carcinogens have been studied [10, 11]. It has also been reported that the phototropism of radish hypocotyls promotes myrosinase gene expression, leading to MTBITC production in the illuminated side of the plant [12, 13]. Increased MTBITC has been shown to induce expression of a heat shock protein that increases the heat resistance of the plant [14]. In shredded cabbage, allyl isothiocyanate (AITC) has been shown to inhibit browning, ethylene production, and respiration [15]. Furthermore, downregulation of phenylalanine ammonia-lyase gene expression by AITC treatment has been observed [16].

However, ITCs are unstable in aqueous solution, and their retainment in processed foods with long shelf lives is difficult. Specifically, the degradation of MTBITC is faster than other ITCs and is completely gone within a few hours in processed *Daikon oroshi* (grated radish) [17, 18].

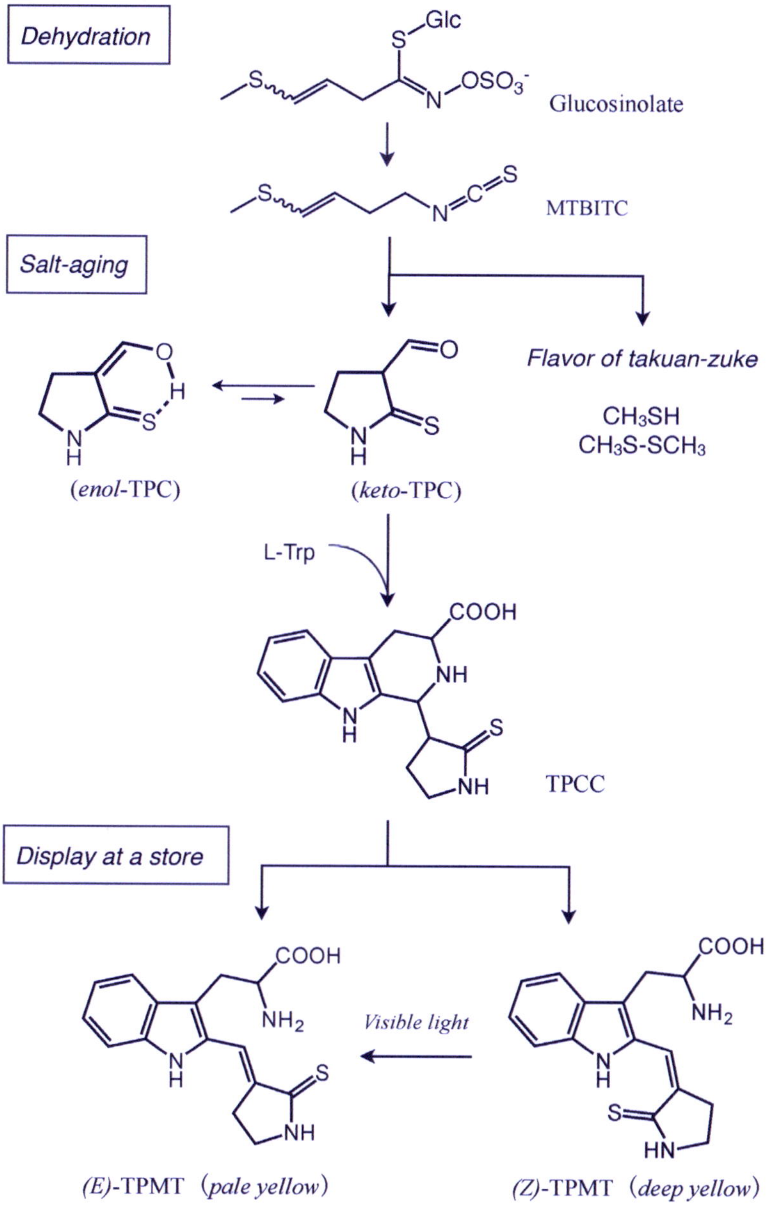

Figure 1.
Pathway of yellowing reaction of radish isothiocyanate.

We present a study of the multistep mechanism of white daikon to yellow takuan-zuke during the process of pickling. This process has shown to induce secondary metabolism in the radish leading to accumulation of hypotensive factor, γ-aminobutyric acid (GABA). We also show that intake of takuan-zuke, containing GABA, is effective in hypertensive model rats. Therefore, our study provides useful information to both consumers and pickle manufacturers.

2. Yellowing mechanism of salted radish root (takuan-zuke)

For takuan-zuke production, dehydration of the raw radish is required as a pre-treatment process. This process destroys the cells and tissues of the plant and activates isothiocyanate production. The yellow-color change of takuan-zuke is the result of a four-step reaction (**Figure 1**).

Step 1: Physical damage to plant cells and tissues by osmotic shock or stone weight activates the myrosinase reaction to generate MTBITC from GRH.

Step 2: MTBITC is converted to 2-thioxo-3-pyrrolidinecarbaldehyde (TPC)-releasing methanethiol due to high reactivity with water molecules [17, 19].

Step 3: The aldehyde group of TPC and the amino group of tryptophan, which is produced by microorganisms or self-maturation in the fermentation process, are condensed by the Pictet-Spengler reaction and converted to 1-(2-thioxopyrrolidine-3-yl)-1,2,3,4-tetrahydro-β-carboline-3-carboxylic acid (TPCC) [2, 20]. This reaction occurs under acidic pH conditions.

Step 4: The ring structure of TPCC is cleaved to form a yellow pigment, 2-[3-(2-thioxopyrrolidin-3-ylidene)methyl]-tryptophan (TPMT). This reaction can occur under either weakly acidic or neutral pH conditions [21].

Steps 1 and 2 occur during the dehydration process. TPC, which is a major degradation product of 4-methylthio-3-butenyl (MTBI), is an important intermediate. In Daikon oroshi, the MTBI produced is rapidly converted to TPC, with the conversion rate reaching 85% after 2 hours [17]. In takuan-zuke, the radish turns yellow, without browning, by passing through this reaction pathway. Therefore, the yellowing reaction is an indicator of the degree of salt-aging.

2.1 Behavior of yellow pigment and related components during the salt-aging process

We prepared sun-dried and salt-pressed takuan-zuke under normal-temperature conditions (0–2 months at 5°C, 2–4 months at 10°C, and 4–8 months at 20°C) and low-temperature conditions (0–8 months at 5°C). The salt-aging conditions of takuan-zuke were as follows: sun-dried normal-temperature (DN) takuan-zuke, sun-dried low-temperature (DL) takuan-zuke, salt-pressed normal-temperature (SN) takuan-zuke, and salt-pressed low-temperature (SL) takuan-zuke. After 8 months, the salinity of the DL and SL samples was 8–9%, and the salinity of DN and SN samples was 15–16%.

Figure 2 shows the yellowing of DN samples. The effect of pH on the TPCC/TPMT yellow-color change reaction of the long-term salt-aging process was analyzed over time (**Figure 3**). The yellow pigment, TPMT, was produced at 9 ± 3 μmol/kg after 4 months of salt-aging and reached 63 μmol/kg after 8 months in the DN sample. Although the other samples did turn yellow, compared to the DN sample, they had smaller amounts of TPMT.

The pH of the DN sample was maintained at ≥5, while the other samples became acidic during the salt-aging treatment. It was suggested that the pH of the sample during salt-aging contributes to the yellowing reaction. However,

Figure 2.
Yellowing of sun-dried normal-temperature takuan-zuke (DN). Zero number indicated sun-dried daikon, other numbers indicate salt-aging time (months).

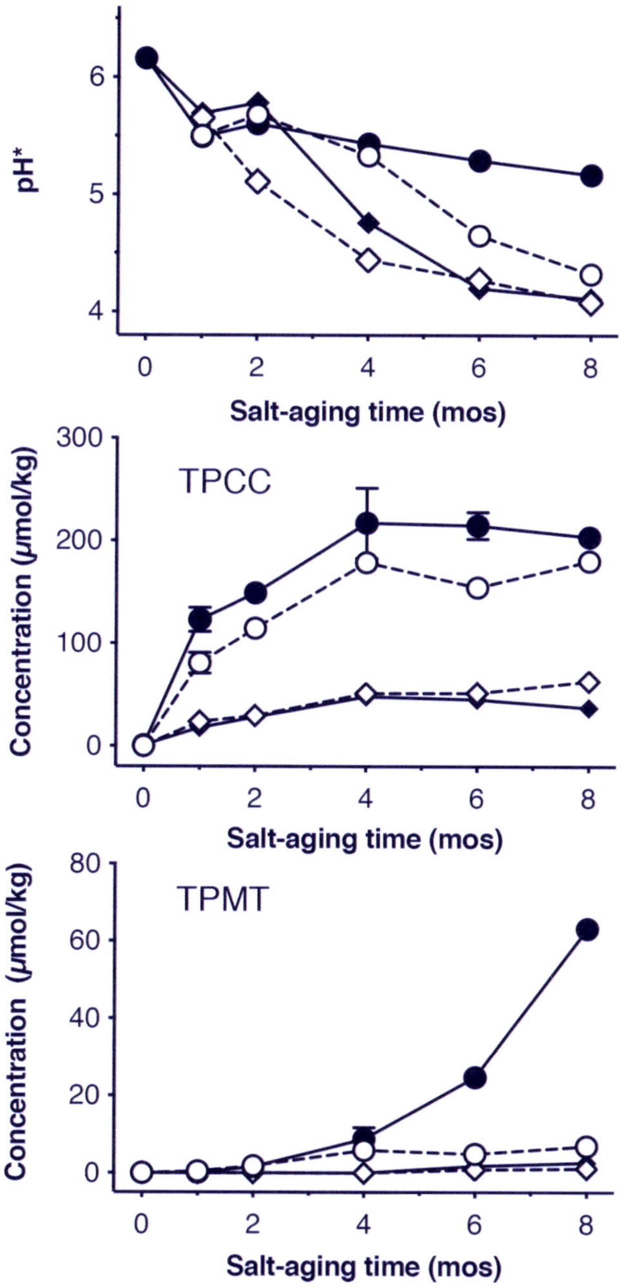

Figure 3.
Change in pH value and concentration of yellow pigment and related components in takuan-zuke. pH value of the brined liquid during the salt-aging process. ◆, salt-pressed normal-temperature takuan-zuke (SN); ◇, salt-pressed low-temperature takuan-zuke (SL); ●, sun-dried normal-temperature takuan-zuke (DN); ○, sun-dried low-temperature takuan-zuke (DL).

it is unclear what role the reaction has on the radish itself. It is reported that TPC has biological functions in humans, including antibacterial activity against food poisoning and cariogenic bacteria and antimutagenic activity against carcinogenic heterocyclic amines [22, 23]. Recently, TPCC and TPMT, which increase with salt-aging, have also been revealed, in our research to have anti-oxidant activities [24], thereby enabling elucidation of the health benefits of takuan-zuke.

3. Metabolomic analysis of salted radish root (takuan-zuke)

Recently, metabolomic analyses involving nuclear magnetic resonance and mass spectrometry have been introduced in the field of food science [25, 26]. Metabolomic analysis enables determination of features by performing comprehensive instrumental and multivariate analyses on a metabolite at a specific moment in time. In the fields of agronomics and food science, it is used for the determination of characteristic secondary metabolites through analysis of variations in different varieties of samples during cultivation, storage, and processing. A mass spectrometer is nonessential, as it is possible to perform multivariate analysis on results obtained from a conventional detector. A description of the application of the latter method to the metabolomic analysis of takuan-zuke can be found in our previously published study [27].

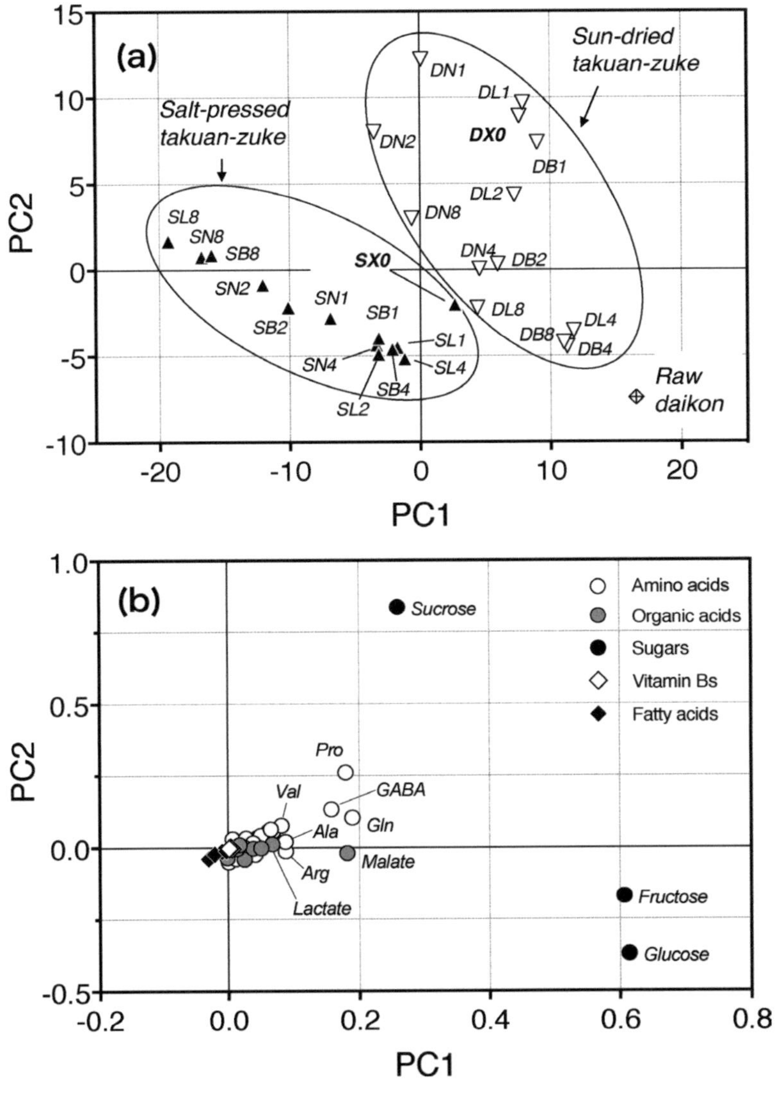

Figure 4.
Multivariate analysis of all takuan-zuke groups. (a) PCA score plots generated using all samples. (b) PCA loading plots from all samples. SN, salt-pressed normal-temperature takuan-zuke; SL, salt-pressed low-temperature takuan-zuke; SB, salt-pressed low-temperature takuan-zuke with rice bran; DN, sun-dried normal-temperature takuan-zuke; DL, sun-dried low-temperature takuan-zuke; DB, dried low-temperature takuan-zuke with rice bran. Numbers indicate salt-aging time (months).

Past research of daikon has focused on its characteristic, pungent flavors. Therefore, we focused on hydrophilic components including amino acids, organic acids, and sugars that were different from those of the pungent flavor. We examined the differences between salt-pressed and sun-dried takuan-zuke, conventional high-salt normal-temperature storage and more recent low-salt low-temperature storage, and the influence of rice bran during salt-aging using the metabolomic analysis method.

We analyzed the components over time in the salt-pressed and sun-dried takuan-zuke for 8 months under the conditions of normal temperature (high-salt), low temperature (low-salt), and low temperature (low-salt, rice bran). In takuan-zuke prepared in 2010, amino acids (glutamine and GABA), organic acid (malic acid), sugars (glucose and fructose), and free fatty acids (α-linolenic, palmitic, and linoleic acid) were detected as major components. Principal component analysis (PCA) shows that the differences in the components of each takuan-zuke depend on the method of dehydration applied (**Figure 4a**). The results of the PCA loading plot show components such as sucrose, proline, and GABA were detected in the first quadrant of **Figure 4b** which corresponds to sun-dried takuan-zuke. In addition, fructose and glucose were detected in the fourth quadrant, which corresponds to raw daikon. However, no characterization of salt-pressed takuan-zuke was found.

Plants under osmotic stressors such as salt and dryness promote the synthesis of proline, a low-toxicity, and compatible solute for water retention [28, 29]. In the

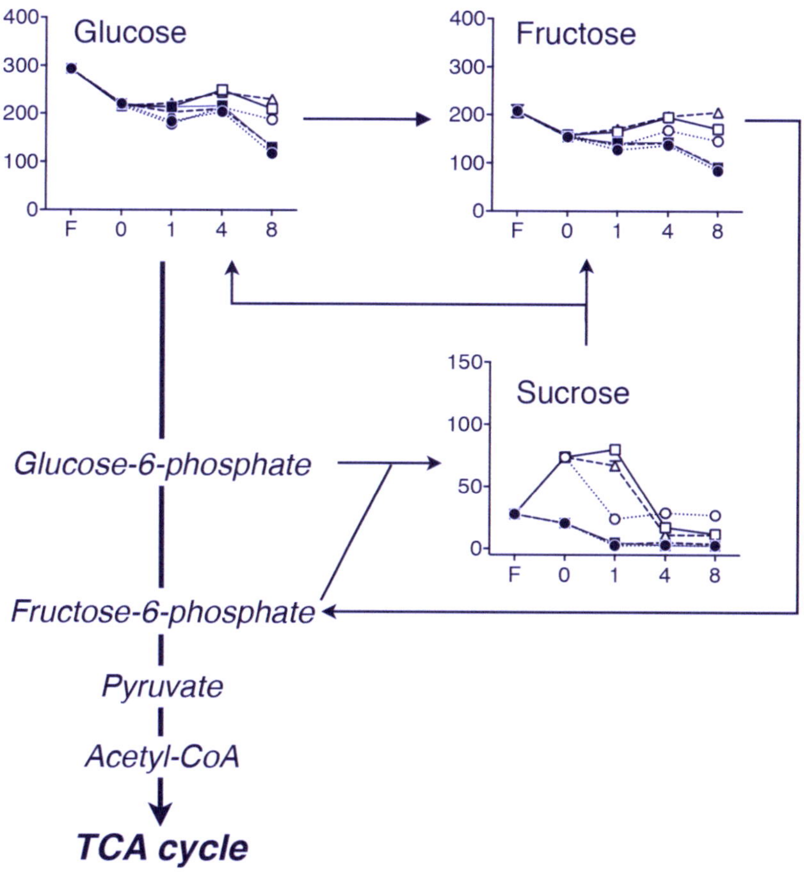

Figure 5.
Time-dependent changes in metabolite concentrations. The x axis denotes processing time (months), and the y axis denotes concentration (mg/g of dry weight, mean ± SD). Zero time denotes the start of salt-aging for dehydration. Symbols denote the following: ○, DN; □, DL; △, DB; ●, SN; ■, SL; ▲, SB.

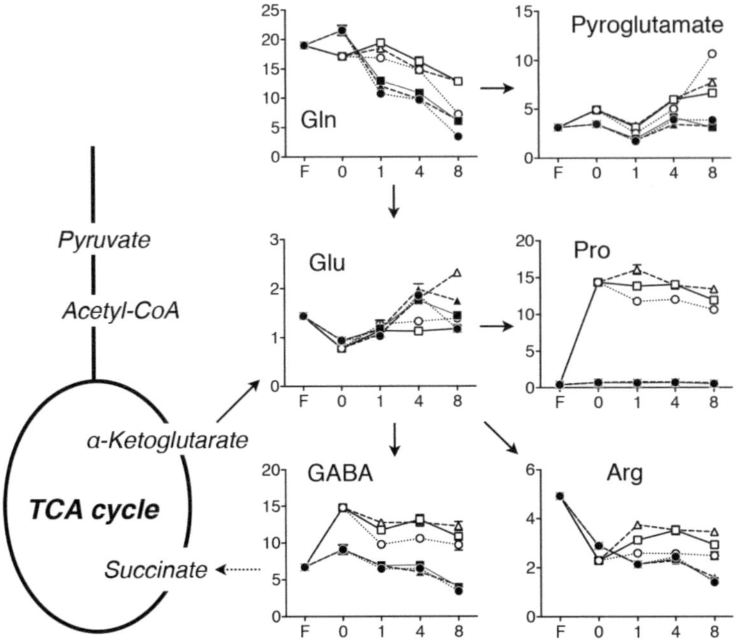

Figure 6.
Time-dependent changes in concentrations of metabolites related to Glu and GABA. The x axis denotes processing time (months), and the y axis denotes concentration (mg/g of dry weight, mean ± SD). Zero time denotes the start of salt-aging for dehydration. Symbols denote the following: ○, DN; □, DL; △, DB; ●, SN; ■, SL; ▲, SB.

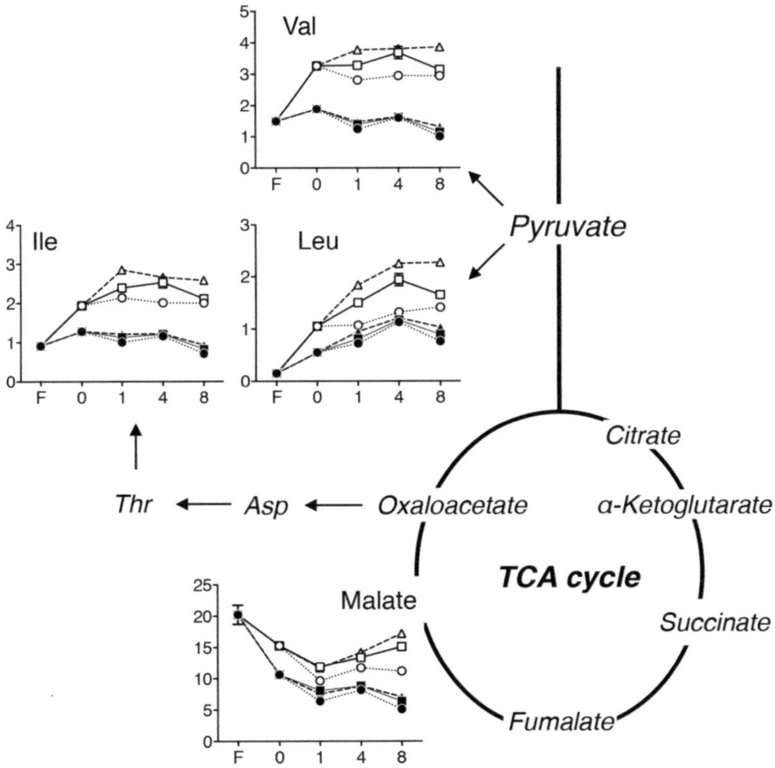

Figure 7.
Time-dependent changes in concentration of metabolites related to BCAA. The x axis denotes processing time (months), and the y axis denotes concentration (mg/g of dry weight, mean ± SD). Zero time denotes the start of salt-aging for dehydration. Symbols denote the following: ○, DN; □, DL; △, DB; ●, SN; ■, SL; ▲, SB.

sun-dried takuan-zuke samples, concentrations of sucrose and proline increased significantly with processing time. In the salt-pressed radish samples, it was suggested that metabolic rates decreased due to the mechanical pressure and osmotic dehydration treatments imposed upon the raw daikon. This dehydration process is called *shio-goroshi* among Japanese manufactures.

Figures 5–7 show analyses of carbohydrate, Glu-GABA, and BCAA composition in the form of metabolic maps. Under both dehydrating conditions, branched-chain amino acids (Val, Leu, Ile), GABA, and other minor components with known functionalities tended to increase with processing time. Furthermore, free polyunsaturated fatty acids and pyroglutamic acid increased with salt-aging time. In salt-aging samples treated with rice bran (SBs and DBs), niacin, glutamic acid, and acetic acid, derived from the rice bran, were found to be dependent on salt-aging time rather than the method of dehydration when under the same temperature conditions (data not shown).

The above reactions are significant for sun-dried takuan-zuke, relating to its yellowing reaction tendency. The concentration of total metabolites in the sun-dried samples is higher than those of salt-pressed takuan-zuke. This shows that in the sun-dried takuan-zuke, it is not the simple effect of drying but the induced secondary metabolic reaction from dehydration that affects the metabolism of the functional components.

4. GABA accumulation during daikon dehydration

Plants accumulate GABA in their cells when they are subjected to physicochemical stressors. This is because the proton-consuming glutamate decarboxylase (GAD) reaction is activated to prevent acidification in cells due to stress, and the pH is simultaneously neutralized with GABA production [30, 31]. This pathway is called "the GABA shunt" and involves synthesis of glutamate from α-ketoglutaric acid, an intermediate of the tricarboxylic acid cycle (TCA), to the synthesis of GABA by the GAD reaction. Following the removal of the stressor, succinic acid is further synthesized by GABA transaminase (GABA-T) and succinic semialdehyde dehydrogenase (SSADH) and enters the TCA cycle. Bown reported that the metabolism of glutamate to succinate via the GABA shunt is energetically less favorable than it is via the TCA cycle [32].

The dehydration treatments imposed by salt and weight are strong stressors on vegetables. As a result, they are convenient for activating GABA synthesis with simultaneous inhibition of GABA metabolism. Since our metabolomic analyses of takuan-zuke (Section 2) demonstrated an increase in GABA production, we further studied Glu-GABA during dehydration [33].

4.1 Effects of dehydration processes on GABA concentration and GAD activity

The GABA content of raw daikon harvested in 2013 was 0.28 ± 0.01 mg/g, which increased to 4.9 ± 0.0 mg/g (DW) with salt-pressing treatment and 9.1 ± 0.1 mg/g with sun-drying treatment; the substrate glutamate decreased (**Figure 8**). In general, the GABA content of Japanese radish differs from harvest year to harvest year due to various factors such as weather, temperature, and soil moisture. Analysis of raw radish from 2012 to 2015 revealed a negative correlation between GABA and the glutamate content ($y = -0.429x + 3.14$, $r^2 = 0.884$). It also seemed that the GABA content fluctuated with postharvest storage conditions. Therefore, all daikon samples were frozen within 8 hours of their harvest.

For the salt-pressed daikon samples, GABA production reached a plateau 2 days after salting. In contrast, in the sun-dried daikon samples, GABA synthesis continued for 3 weeks. The pH values of the sun-dried daikon samples remained between 6.0 and 6.5, while those of the salt-pressed daikon samples decreased over time.

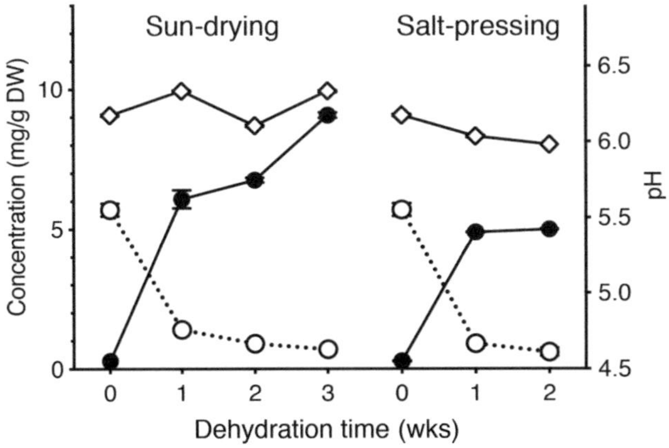

Figure 8.
Time-dependent changes of GABA, Glu, and pH during dehydration process. Symbols denote the following: ◇, pH; ○, Glu; ●, GABA.

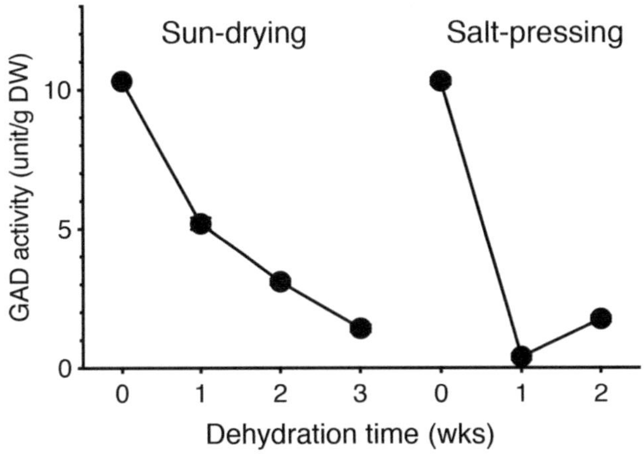

Figure 9.
Time-dependent changes of GABA, Glu, and pH during dehydration process.

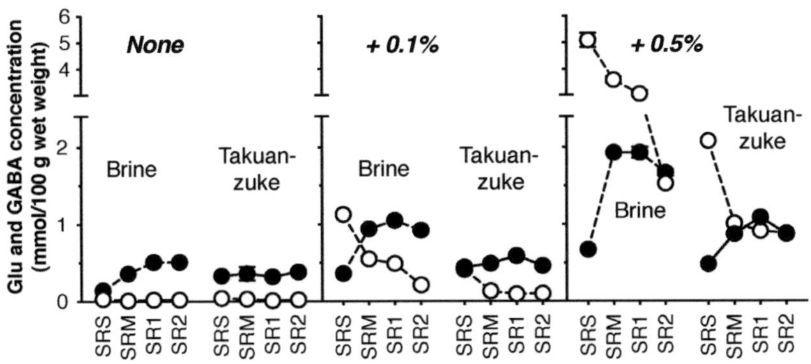

Figure 10.
Effect of monosodium glutamate on GABA concentration in the salt-pressed takuan-zuke and the brine during dehydrating and salting processes. The symbols denote the following: ○, Glu; ●, GABA.

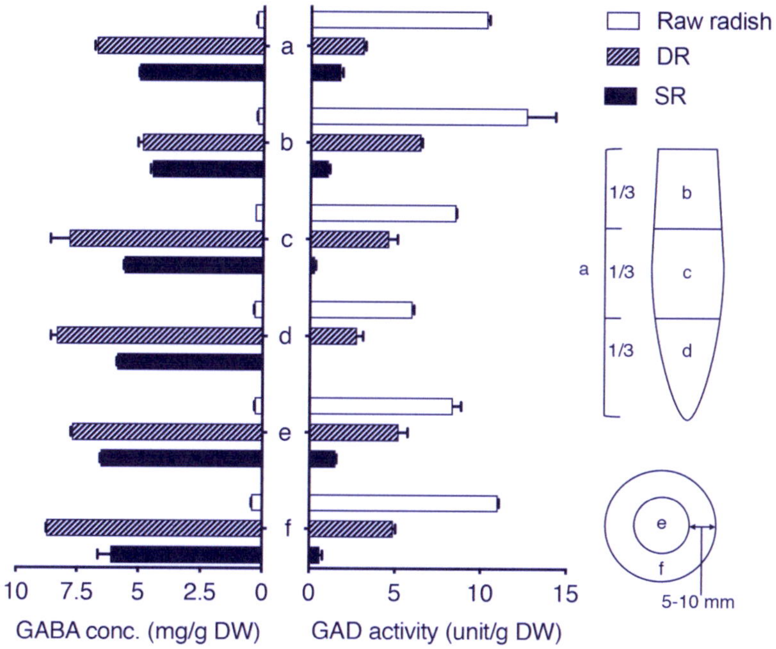

Figure 11.
Distribution of GABA and GAD in dehydrated radish roots. DR, sun-dried radish after 2 weeks; SR, salt-pressed radish after 2 weeks.

Regarding GAD activity, which is involved in GABA synthesis and pH control, the fresh radish had the highest activity, and activity decreased with dehydration treatment time. The enzyme activity decreased gradually with the sun-drying treatment, while it decreased rapidly with the salt-pressing treatment (**Figure 9**). From these results, it was suggested that the glutamate biosynthesis pathway of glutamine is maintained during sun-drying and the GABA synthesis reaction is maintained. It was also revealed that the addition of high amounts of salt inactivated GAD activity, reducing its pH control function.

We prepared GABA-enriched takuan-zuke with the addition of monosodium glutamate (MSG) at the start of salting. As a result, not only GABA concentration in the salted radish increased but also that in the brine (**Figure 10**). The penetration rate of MSG into the radish root is slower than that of NaCl. Thus, GABA, which was converted from MSG in the brine, penetrated into the radish root depending on the concentration gradient. Ueno et al. isolated high-GABA-producing lactic acid bacteria from Senmaizuke, a traditional Kyoto pickle [34]. In takuan-zuke, it is expected that a high-GABA-producing lactic acid bacterium is activated by MSG addition.

4.2 Distribution of GABA concentration and GAD activity in dehydrated radish roots

The distribution of GAD in radish samples was relatively high in the upper root portions, including the base and outer vascular cambium (**Figure 11**). There was no correlation found between residual GAD activity and GABA production. It has been reported that myrosinase, an enzyme involved in MTBITC synthesis, is localized to the epidermis and cambium, and its activity ratio is reported to be 19-fold [35]. The lack of GAD localization showed that the system responding to environmental stressors is functional entirely at the base and root.

5. Conclusion

Takuan-zuke, which is a fermented food, changes the microflora such as lactic acid bacteria and yeast during salt-aging. However, it was not possible to find a dynamic chemical change contributing to microbial fermentation, because the chemical changes derived from the endogenous radish component are large. Kato reported that the presence of long-term-salt-aging-activated *Debaryomyces hansenii*, which is a halo-tolerant yeast; consumption of lactic acid; and generation of ammonia from amino acid during fermentation suppress the decrease in pH [36]. This is presumed to be the reason why the pH of sun-dried normal-temperature takuan-zuke (DN samples) was maintained. The pH condition was optimal for the yellowing reaction.

Recently, taste research studies have revealed that GABA, which in the past had been considered tasteless, affects other tastes. It has also been shown that GABA has an enhancement effect on salty taste [37, 38]. We reported that the intake of takuan-zuke, containing GABA antihypertensive factor, improves renal function and suppresses blood pressure elevation in hypertension model rats [39]. The report also revealed that antihypertensive effects are high in salt-pressed takuan-zuke, which has low GABA content. These results suggest the presence of another antihypertensive factor, different from GABA. In particular, GABA production by simple pickle processing seems to be the key to developing future pickles in terms of both taste function and health benefits.

Acknowledgements

We are grateful to Prof. Emeritus Y. Uda and Prof. Yoichi Yamada of Utsunomiya University, Prof. Yoshio Ozawa of Takasaki University of Health and Welfare, and Prof. Emeritus T. Iizuka of Gunma University. We thank the many students in our laboratory. This work was supported in part by a Sasakawa Scientific Research Grant from The Japan Science Society. We would like to thank Editage (www.editage.jp) for English language editing.

Conflict of interest

The authors declare no conflict of interest.

Author details

Hiroki Matsuoka[1*], Kei Kumakura[1], Taito Kobayashi[1], Wataru Kobayashi[1] and Asaka Takahashi[2]

1 Department of Health and Nutrition, Takasaki University of Health and Welfare, Takasaki, Gunma, Japan

2 Faculty of Nutritional Sciences, Tohto University, Fukaya, Saitama, Japan

*Address all correspondence to: matsuoka@takasaki-u.ac.jp

References

[1] Kjær A, Madsen JO, Maeda Y, Ozawa Y, Uda Y. Volatiles in distillates of fresh radish of Japanese and Kenyan origin. Agricultural and Biological Chemistry. 1978;**42**(9):1715-1721

[2] Ozawa Y, Uda Y, Kawakishi S. Generation of β-carboline derivative, the yellowish precursor of processed radish roots, from 4-methylthio-3-butenyl isothiocyanate and L-tryptophan. Agricultural and Biological Chemistry. 1990;**54**(7):1849-1851

[3] Ozawa Y, Uda Y, Ohshima T, Saito K, Maeda Y. Formation of yellow pigment by the reaction of 4-methylthio-3-butenyl isothiocyanate with L-ascorbic acid and some dihydroxyphenolic compounds. Agricultural and Biological Chemistry. 1990;**54**(3):605-611

[4] Ozawa Y, Uda Y, Kawakishi S. Formation of yellow pigment containing a β-carboline skeleton in salted radish roots and its properties. Nippon Shokuhin Kogyo Gakkaishi (in Japanese). 1992;**39**(6):490-495

[5] Nakamura Y, Nakamura K, Asa Y, Wada T, Tanaka K, Matsuo T, et al. Comparison of the glucosinolate–myrosinase systems among daikon (*Raphanus sativus*, Japanese white radish) varieties. Journal of Agricultural and Food Chemistry. 2008;**56**(8):2702-2707

[6] Goi H, Inoue S, Iwanami Y. Antifungal activity of powdery black mustard, powdery wasabi (Japanese horseradish), and allyl isothiocyanate by gaseous contact—Antifungal activity of plant volatiles. Journal of Antibacterial and Antifungal Agents. 1985;**13**(5):199-204

[7] Drobnica L, Zemanova M, Nemec P, Antos K, Kristian P, Stullerova A, et al. Antifungal activity of isothiocyanates and related compounds. Applied Microbiology. 1967;**15**(4):701-709

[8] Agerbirk N, Olsen CE. Glucosinolate structures in evolution. Phytochemistry. 2012;**77**:16-45

[9] Kanemaru K, Miyamoto T. Inhibitory effects on the growth of several bacteria by brown mustard and allyl isothiocyanate. Nippon Shokuhin Kagaku Kogaku Kaishi. 1990;**37**(10):823-829

[10] Nakamura Y, Miyoshi N. Electrophiles in foods: The current status of isothiocyanates and their chemical biology. Bioscience, Biotechnology, and Biochemistry. 2010;**74**(2):242-255

[11] Nakamura Y, Morimitsu Y, Uzu T, Ohigashi H, Murakami A, Naito Y, et al. A glutathione S-transferase inducer from papaya, rapid screening, identification and structure-activity relationship of isothiocyanates. Cancer Letters. 2000;**157**(2):193-200

[12] Hasegawa T, Yamada K, Kosemura S, Yamamura S, Hasegawa K. Phototropic stimulation induces the conversion of glucosinolate to phototropism-regulating substances of radish hypocotyls. Phytochemistry. 2000;**54**(3):275-279

[13] Yamada K, Hasegawa T, Minami E, Shibuya N, Kosemura S, Yamaura S, et al. Induction of myrosinase gene expression and myrosinase activity in radish hypocotyls by phototropic stimulation. Journal of Plant Physiology. 2003;**160**:255-259

[14] Hara M, Harazaki A, Tabata K. Administration of isothiocyanates enhances heat tolerance in *Arabidopsis thaliana*. Plant Growth Regulation. 2013;**69**(1):71-77

[15] Yano M, Saijo R, Ota Y. Inhibition of browning and ethylene

production in shredded cabbage by isothiocyanates. Journal of the Japanese Society for Horticultural Science. 1986;**55**(2):194-198

[16] Banerjee A, Penna S, Variyar PS. Allyl isothiocyanate enhances shelf life of minimally processed shredded cabbage. Food Chemistry. 2015;**183**:265-272

[17] Kobayashi T, Kumakura K, Kobayashi W, Takahashi A, Matsuoka H. Determination of 2-thioxo-3-pyrrolidinecarbaldehyde in salted radish root (takuan-zuke) by high-performance liquid chromatography with fluorescence detection after pre-column derivatization using 4-(*N,N*-dimethylaminosulfonyl)-7-hydrazino-2,1,3-benzoxadiazole. Separations. 2017;**4**(4):35

[18] Uda Y, Matsuoka H, Kumagami H, Shima H, Maeda Y. Stability and antimicrobial property of 4-methylthio-3-butenyl isothiocyanate, the pungent principle in radish. Nippon Shokuhin Kogyo Gakkaishi. 1993;**40**(10):743-746

[19] Uda Y, Ozawa Y, Ohshima T, Kawakishi S. Identification of enolated 2-thioxo-3-pyrrolidinecarbardehyde, a new degradation product of 4-methylthio-3-butenyl isothiocyanate. Agricultural and Biological Chemistry. 1990;**54**(3):613-617

[20] Ozawa Y, Kawakishi S, Uda Y, Maeda Y. Isolation and identification of a novel β-carboline derivative in salted radish roots, *Raphanus sativus* L. Agricultural and Biological Chemistry. 1990;**54**(5):1241-1245

[21] Matsuoka H, Takahashi A, Ozawa Y, Yamada Y, Uda Y, Kawakishi S. 2-[3-(2-thioxopyrrolidin-3-ylidene) methyl]-tryptophan, a novel yellow pigment in salted radish roots. Bioscience, Biotechnology, and Biochemistry. 2002;**66**(7):1450-1454

[22] Uda Y, Hayashi H, Takahashi A, Shimizu A. Mutagenic and antimutagenic property of 3-hydroxymethylene-2-thioxopyrrolidine, a major product generating from pungent principle of radish. Food Science and Technology—Lebensmittel—Wissenschaft & Technologie. 2000;**33**(1):37-43

[23] Hashimoto K, Yanagi K, Fukushima K, Uda Y. Effect of 3-hydroxymethylene-2-thioxopyrrolidine on growth of two species of mutans streptococci and their in vitro plaque formation. International Journal of Antimicrobial Agents. 2001;**17**(2):97-102

[24] Takahashi A, Matsuoka H, Watanabe E, Shinagawa A, Ozawa Y, Yamada Y, et al. Quantitative analysis of yellow pigment in takuan-zuke and their ABTS radical cation scavenging activity. Food Science and Technology Research. 2009;**15**(3):337-342

[25] Ko BK, Ahn HJ, van den Berg F, Lee CH, Hong YS. Metabolomic insight into soy sauce through (1) H NMR spectroscopy. Journal of Agricultural and Food Chemistry. 2009;**57**(15):6862-6870

[26] Sugimoto M, Goto H, Otomo K, Ito M, Onuma H, Suzuki A, et al. Metabolomic profiles and sensory attributes of edamame under various storage duration and temperature conditions. Journal of Agricultural and Food Chemistry. 2010;**58**:8418-8425

[27] Kumakura K, Kato R, Kobayashi T, Sekiguchi A, Kimura N, Takahashi H, et al. Nutritional content and health benefits of sun-dried and salt-aged radish (*takuan-zuke*). Food Chemistry. 2017;**231**:33-41

[28] Liu C, Zhao L, Yu G. The dominant glutamic acid metabolic flux to produce γ-amino butyric acid over proline in *Nicotiana tabacum*

leaves under water stress relates to its significant role in antioxidant activity. Journal of Integrative Plant Biology. 2011;**53**(8):608-618

[29] Silveira JAG, de Almeida Viégas R, da Rocha IMA, ACdOM M, de Azevedo Moreira R, JTA O. Proline accumulation and glutamine synthetase activity are increased by salt-induced proteolysis in cashew leaves. Journal of Plant Physiology. 2003;**160**(2):115-123

[30] Streeter JG, Thompson JF. Anaerobic accumulation of γ-aminobutyric acid and alanine in radish leaves (*Raphanus sativus* L.). Plant Physiology. 1972;**49**:572-578

[31] Streeter JG, Thompson JF. In vivo and in vitro studies on γ-aminobutyric acid metabolism with the radish plant (*Raphanus sativus* L.). Plant Physiology. 1972;**49**:579-584

[32] Bown AW, Shelp BJ. The metabolism and Functions of y-Aminobutyric Acid. Plant Physiology. 1997;**115**(1-5):1-5

[33] Kato R, Hayashi S, Kobayashi T, Takahashi H, Kimura N, Takahashi A, et al. Behavior analysis of γ-aminobutyrate and glutamate decarboxylase activity in salted radish roots (*takuan-zuke*). Nippon Shokuhin Kagaku Kogaku Kaishi (in Japanese). 2015;**62**(10):492-500

[34] Ueno Y, Hiraga K, Mori Y, Oda K, et al. Isolation and utilization of a lactic acid bacterium, producing a high level of γ-aminobutyric acid (GABA). Seibutsu Kogaku Kaishi. 2007;**85**(3):109-114

[35] Hara M, Etoh H, Kuboi T. Tissue printing for myrosinase activity in roots of turnip and Japanese radish and horseradish: A technique for localizing myrosinase. Plant Science. 2001;**160**:425-431

[36] Kato S, Kitamura E, Ohshima S. Factors influencing the growth of *Debaryomyces hansenii* and *Saccharomyces servazii* isolated from salted "Daikon" (Japanese radish). Nippon Shokuhin Kogyo Gakkaishi. 1991;**38**(4):357-359

[37] Sasaki K, Hamano K, Ueno H. Relationship between GAD/GABA and sweet, umami and bitter tastes, those received in type II taste buds. Journal of Biological Macromolecules. 2014;**14**(1):43-57

[38] Ueno H. Development of a quantitative screening method for the salty taste enhancement effect of plant-derived substances: The potential application for use in flavor enhancement of salt reduced food products. Kagaku to Seibutsu. 2014;**52**(11):742-748

[39] Kumakura K, Kato R, Kobayashi T, Kimura N, Takahashi H, Takahashi A, et al. The salted radish takuan-zuke shows antihypertension effects in spontaneously hypertensive rats. Food & Function. 2017;**8**(10):3491-3500